CROWDED ORBITS

CROWDED ORBITS
Conflict and Cooperation in Space

JAMES CLAY MOLTZ

Columbia University Press New York

Columbia University Press
Publishers Since 1893
New York   Chichester, West Sussex
cup.columbia.edu
Copyright © 2014 Columbia University Press
All rights reserved
Library of Congress Cataloging-in-Publication Data
Moltz, James Clay.
Crowded orbits : conflict and cooperation in space / James Clay Moltz.
      pages   cm
Includes bibliographical references and index.
isbn 978-0-231-15912-8 (cloth : alk. paper) — isbn 978-0-231-52817-7 (electronic)
1. Outer space—Exploration. 2. Planets—Exploration. 3. Astronautics and state.
4. Astronautics—International cooperation. 5. Space law. 6. Space security.
I. Title.

QB500.25M67   2014
629.4'1—dc23
2013028213

Columbia University Press books are printed on permanent and durable acid-
free paper.
This book is printed on paper with recycled content.
Printed in the United States of America
c 10 9 8 7 6 5 4 3 2 1

Jacket design: Chris Sergio

References to websites (URLs) were accurate at the time of writing. Neither the
author nor Columbia University Press is responsible for URLs that may have
expired or changed since the manuscript was prepared.

CONTENTS

PREFACE

This book offers general readers and students an understanding of the competing trends of competition and cooperation in the past and present of human space activity, while also asking questions about the future. It covers scientific topics, the economics of space, and difficult debates about military security. The book does not assume any prior knowledge of space and is aimed at reaching anyone with an interest in the subject matter. Its overall focus is on possible means of avoiding dangerous conflict in this new environment, even as human activity increases.

I am grateful to Columbia University Press, especially Anne Routon, for taking an interest in this topic. The approach adopted here is similar to Joseph Cirincione's in *Bomb Scare: The History and Future of Nuclear Weapons* (Columbia University Press, 2007), an engaging and informative short volume on a key policy issue facing nations in a crucial arena of international relations, including the risks of future conflict. I have tried to give readers a full picture of the historical, technological, and political factors in space necessary to allow them to discuss and analyze the issues intelligently.

I wrote this volume over the course of the three years from 2011 to 2013. But the book also draws on my more than twenty-five years of professional experience studying space politics and writing about their international dimension. While there is considerable academic literature on space competition and cooperation, almost all of it is written in scholarly and specialist jargon. This volume is an effort to bring the key concepts and problems to the interested public and students, who have to date been underserved in terms of accessible studies on international space policy.

For their generous willingness to read my draft chapters and their many useful comments and corrections, I thank several of my colleagues in the Naval Postgraduate School's Space Systems Academic Group: Karen Andersen, Commander (U.S. Navy, Retired) Charles Racoosin, Dr. Jennifer Rhatigan, and Captain (U.S. Navy, Retired) Alan Scott. Their expert suggestions on specific chapters have strengthened the book considerably. I also thank my wife, Sarah J. Diehl, whose incisive grammatical and substantive critique on various chapters helped improve the book's readability. Finally, I am grateful to two anonymous readers who reviewed the book for Columbia University Press and offered very useful comments.

Despite the strength of all this expert help, I remain responsible for any errors or oversights in this book. In addition, all opinions expressed here are my own, and should not be interpreted as the official policies of the U.S. Navy or the Department of Defense. Nevertheless, it is my firm hope that readers will come away better educated on this subject and, just as importantly, interested in learning more.

CROWDED ORBITS

INTRODUCTION

Competing nations have thus far managed to avoid direct conflict in space. Given past battles over land territories, on the world's oceans, and in the air, the record of humans in space since the first satellite launch in 1957 is impressive. But will countries be able to keep the peace as space becomes more crowded? This is a simple and yet very important question that requires greater attention.

In the 1960s television program *Star Trek*, the countries of the world finally, by the twenty-second century, develop a cooperative organization for working together in space, the United Federation of Planets. It includes all people on Earth and also beings from friendly planets. But the route to the federation, as described in the program, was a very costly one: nuclear war on Earth, battles in space, and only an ex post facto recognition that the nations on Earth had fundamentally *shared* interests as they explored the galaxies. Like the rapprochements in real life that ended the prior divisions within Western Europe after 1945 and between Western and Eastern Europe after 1989, this basic revelation had taken many years of conflict to be realized. In the *Star Trek* television series, international cooperation comes about only under the threat of the end of human civilization—almost as a last resort. If the nuclear wars that took place had gone worse, this cooperative escape route might have been snuffed out entirely. This fictional metaphor is hardly a positive one for the coming generations of people on this planet or for the next few hundred years in space. Can we do better? If so, how?

The risk of space conflict raises a number of troubling challenges as we stand on the threshold of a major expansion of human space activity. In order to prevent space warfare, we will need to understand the preconditions for bringing about greater collaboration among Earth's nations in orbit. Can we achieve cooperation sooner than in the *Star Trek* series and without having to go through the possibly disastrous effects of nuclear or space war? Might successful cooperation in near-Earth space—such as on the *International Space Station* (*ISS*)—serve as a first step? Perhaps, but the current space station does not include a number of important new spacefaring countries. Also, recent destructive activities in low-Earth orbit by China and the United States and threats by other nations to develop similar anti-satellite capabilities presage difficulties in overcoming international mistrust in space.

Yet humans have an amazing potential to learn and to engage in self-restraint, once they figure out that it is in their best interests to do so. If there is a single lesson from the Cold War in space, it is that both sides eventually learned that unrestricted military behavior risked uncontrollable conflict and the possible ruination of the near-Earth space environment, thereby *worsening* their individual and mutual security. For this reason, Washington and Moscow exercised remarkable self-restraint even during the most hostile years of the space race, forged non-interference agreements, and never fired shots in anger in space. Ironically, in some respects we seem further from such cooperative policies in the early twenty-first century.

Achieving a peaceful and sustainable international approach to space will require an even firmer commitment to responsible behavior among today's emerging space actors because useful orbits are becoming more populated than ever before. What each country or other spacefaring entity does has the potential to affect everyone else. But can we shift from a traditional focus on self-interest and beating our adversaries to a focus on broader goals for humankind, such as peaceful development, joint policing, and collaborative colonization of space? Perhaps if we step back to recognize that we exist in a rough universe for human life—at least judging by the apparent rarity of our species among the known planets—we can begin to broaden our perspective of what is truly important.

But many expert predictions are pessimistic on this score. The current debate about future space policy encompasses three basic points of view. From least to most cooperative, first, there are those who believe that space will inevitably become a struggle for military hegemony by one nation or a group of nations, despite the dangers posed by space warfare. They chalk it up to human nature, which they posit as imperfect, and to international mistrust, which they predict is likely to continue, despite economic globalization and the breadth of Internet data-sharing. Second, there are those who believe that harmful space conflicts might be avoided by some form of piecemeal global engagement, or the "muddling through" approach that has worked in the past. Some dangerous activities might occur, but self-interest will likely drive countries to restrain themselves enough not to ruin space altogether. Finally, at the far end of the spectrum, there are those who foresee the prospect of international space governance through new global institutions. They predict that new conditions of increasing international space activity, growing costs for strictly national programs, and improved communications will create "demand" for more comprehensive space organizations and more institutionalized forms of cooperation. Such an evolution might move us toward more of a "humankind" approach to space, rather than one based on competing nations, as in the past. But even if advocates of this last perspective are right, the road to get there could well be a very rocky one. Will that be good enough to preserve safe and productive access to space for the future?

The United States issued a National Security Space Strategy in 2011 that described the emerging space environment as "congested, contested, and competitive." This is the reality that we have to deal with in the twenty-first century as more actors enter Earth orbital space. Countries unanimously say that they are pursuing "peaceful purposes" in space. But will they practice what they preach? The nature of future relations among actors in space is not predetermined and could either become a hostile competition or be characterized by joint development. The outcome of this ongoing mix of factors will be a choice made by the actors themselves. It will also be affected by whether the major participants can form new governance mechanisms for space and how effective these measures will prove to be in practice.

Sadly, the previous default behavior for humans with regard to new environments (unsettled continents, the oceans, and the world's air space) has been a self-destructive pattern of conflict and division of spoils, raising questions about whether human civilization has really advanced. In space, however, the conflicts could be quite a bit more dangerous, particularly given the close linkages between space security and nuclear stability among the great powers. Even limited conflict in space could lead to the possible loss of the near-Earth orbital region because of the release of harmful debris. Moreover, if nuclear weapons are used, highly damaging electromagnetic pulse radiation would remain in orbit for decades. In popular space movies like the *Star Wars* series, spaceships destroyed in orbital battles are typically blown into pieces and then "disappear" into the ether. Why not show audiences the real effects of space warfare? As portrayed in the film *Gravity*, a single explosion in Earth orbital space could create thousands of small fragments that would continue to hurtle around our planet at more than 17,000 miles per hour, without any human control. Space, unfortunately, is not a self-cleaning environment, except over very long periods of time (this means centuries). Thus it is imperative that we learn "fast enough" to prevent such destructive space conflicts.

Figuring out how to get from our current situation to a more cooperative one requires knowledge. We have to do better than the designer of the video game *Star Wars: The Force Unleashed II*, who states about one of its main objectives: "We really wanted to make sure you were encouraged to destroy the world around you."[1] Young people, in particular, need to understand why such fun carries not only obvious moral implications but also self-defeating operational ones in space, such as creating orbital debris that could quickly destroy your own spacecraft.

Developing responsible space policy requires answering a series of difficult questions, as well as undertaking a careful examination of lessons from the past (both in space and in other environments). It also calls upon interested people and government officials to review existing trends in space activities, politics, and technology. This book covers these topics and provides readers with the tools they need to analyze current and emerging issues in space policy. It assumes no particular background in space science, history, or politics, but provides a foundation in each of these areas so that the reader can approach complex subjects regarding the future in space with new knowledge and understanding.

This learning process begins with the basics of space physics and orbital mechanics. Space poses many challenges because of its unique characteristics. It lacks atmospheric pressure, objects traveling through it experience alternating extremes of hot and cold (depending on their relation to the Sun), and spacecraft require fuel sources in order to move. All of these characteristics profoundly affect what can (and cannot) be done in this environment. Far too often, pundits, political officials, and even military leaders have made bold claims about what "will" happen in space, only to be proven wrong by scientists and engineers. That is why understanding these core principles is important.

Also critical to understanding the future of space is the history of what humans have done in this environment, some of it little known to the average citizen. For example, few people know that the United States and the Soviet Union tested nuclear weapons in orbit early in the space age, nearly halting the development of satellite communications and preventing further progress in human spaceflight. Only a treaty between the two superpowers averted what could have been a true dead end for human activity in space. Following the end of the Cold War space race, remarkable U.S.-Russian space cooperation emerged in the 1990s, particularly in joint launch ventures and in building the *ISS* with the European Space Agency (ESA), Canada, and Japan. This recent history also involves the rise of new space actors such as China, India, Iran, Israel, North Korea, and South Korea. Indeed, there are now more than sixty countries and government consortia that own or operate satellites in orbit. The number of yearly launches could jump from fewer than one hundred to as many as a thousand by 2020, once new launch services to both orbital and suborbital space begin. What is unclear is whether this increasing space activity will lead to conflict, or whether new international management mechanisms can prevent it.

Space activity can be divided into several different functional areas. Space science and human exploration of space are often grouped together in the category of "civil" space activity: those projects managed by governments that serve nonmilitary and noncommercial purposes. Before 2000, the Soviet Union/Russia and the United States dominated this field, launching all of the spacecraft that bore human beings and carrying out all but a handful of the missions into deep space (that is, the region beyond Earth and the Moon). The Soviet Union accomplished critical "firsts" in orbital satellites, human spaceflight, space stations, and long-duration spaceflights, and succeeded in a number of planetary missions, in particular to Venus. Key U.S. accomplishments included the first weather, remote-sensing, and communications satellites; the 1969 Moon landing; the massive *Skylab* space station in the early 1970s; the operation of the reusable space shuttles; and exploratory missions such as the *Pioneer, Mariner, Viking, Magellan, Galileo,* and *Ulysses* spacecraft to Venus, Mars, Jupiter, Saturn, Uranus, and Neptune and other celestial bodies. NASA continues this legacy but is facing new budgetary pressures. France and Germany have led other successful deep-space missions. In the past decade, Japan, China, and India have joined these efforts in launching unmanned probes that have landed on asteroids, orbited the Moon, and conducted lunar mapping or other surveying. In human spaceflight, China's successful launch program—beginning in 2003—has stimulated a number of other countries, among them India, Japan, Iran, and the European Space Agency, to pledge or consider independent human spaceflight programs of their own. The continuing expansion of countries involved in civil space activity heightens chances for both future competition and new forms of cooperation.

Commercial space activity is also expanding rapidly in the early twenty-first century. While initial U.S. developments in satellite technology spawned the emergence of the space communications industry in the 1960s, a range of new actors is now populating this sector. In addition, since the 1980s, traditional voice, data, and video communications have been supplemented by new services, such as those made possible by the commercialization of U.S. precision navigation and timing technology (through the Global Positioning System [GPS]), the commercial marketing of various Earth-imaging systems, and the advent of direct broadcasting satellites for television, radio, and Internet broadband. But the growth of space commerce has created certain problems, especially in regard to the finite number of available locations in high-demand regions of space and to the limited number of broadcasting frequencies. Cases of national governments jamming commercial communications signals are increasing, raising fears of service denials. This means that international organizations responsible for managing the space broadcasting realm will have to become more active in enforcing existing rules, protecting law-abiding operators, and sanctioning violators, if space commerce is to remain safe, reliable, and profitable.

Another worry among many observers is the spread of military space technology. Some of these new defense capabilities have a positive dimension, particularly when they contribute to finding terrorists, directing precision weapons against evildoers (and avoiding innocents), and enforcing arms control and nonproliferation treaties. Space-based communications, meteorology, imagery, and signals intelligence systems have proven to be important "enablers" for the U.S. military. But potential kinetic, microwave, laser, and other weapons systems could put these assets at risk, given the transparent nature of space. To counter such threats, national militaries will have to undertake new efforts to reduce satellite vulnerability by building more spacecraft, having the ability to replace them quickly, or loading them with more fuel so that they will be able to avoid attackers. A number of leading spacefaring nations are also considering active defenses. China's 2007 test of a kinetic anti-satellite weapon against one of its own aging satellites highlighted the absence of clear rules with regard to weapons systems that do not employ weapons of mass destruction (which are banned by a treaty). Many weapons technologies can still be legally tested and deployed, which raises the risk of future conflicts as well as the spread of harmful debris.

Space diplomacy is a field that has received comparatively little attention since the 1970s, despite the recent rise of multilateral space tensions. The UN body responsible for the negotiation of new security-related treaties is the Conference on Disarmament (CD), in Geneva. Unfortunately, action by this organization has been frozen for more than fifteen years. From the late 1990s to 2009, a U.S.-Chinese dispute over nuclear versus space arms control priorities blocked agreement on an official agenda. Successful U.S.-Chinese compromises finally promised new action, only to see hopes stymied by Pakistani intransigence on a fissile material production ban, which has prevented any formal negotiations once again. During the Cold War, U.S.-Soviet preeminence in space made it possible for bilateral arms control treaties to resolve the most pressing concerns. Today, the realm of relevant actors is larger and more complex, and less-developed countries are highly reluctant to see new treaties lock in advantages for more-advanced militaries. In addition, political disagreements and mistrust between the United States and China, exacerbated by conservative actors in both countries' domestic politics, have prevented bilateral actions that might help kick-start a broader process. These conditions caused European countries to offer a proposal for a voluntary space code of conduct, which is now under consideration internationally. But some experts believe more substantial agreements will be needed: formal treaties with international monitoring. The questions of who will lead these efforts, provide the necessary space systems to support them, and fund the verification organizations needed to enforce them remain to be answered.

Overall, determining where we are headed in space requires addressing a series of complex questions, among them:

- Will countries collaborate or compete in returning to the Moon and eventually going to Mars?
- Will forces of economic globalization prove more powerful than those of nationalism in motivating military space activity?
- Will space actors be able to collaborate effectively to rein in the hazards posed by orbital debris and other forms of crowding in space?

There are at least three possible scenarios for space activity, depending on the answers to these questions: military control by one nation or a group of countries; issue-specific problem-solving by different groups of interested space actors (muddling through); or expanded international treaties and cooperative institutions.

History suggests that future competition among nations is inevitable and could have beneficial effects in stimulating space innovation. But excess competition could lead to self-destructive conflict. The countries in the *Star Trek* series learned this the hard way. Will the actual twenty-first century in space fulfill this dangerous prediction or accomplish something better? Everyone would agree that we have shared interests in protecting human civilization and expanding into the solar system and beyond. The challenge is how to get from here to there. To start, we need to understand better what has taken place in space thus far.

1

GETTING INTO ORBIT

> Rocket and automobile engines ... have a basic similarity: Both are internal-combustion engines using the burning of a fuel-oxygen mixture to produce hot gases which create tremendous pressure. ... In the automobile, the gases push a piston which ... eventually pushes a wheel against the ground. In the rocket, the gases push ... directly on the vehicle itself ... making the rocket, in effect, a single, huge piston. ... The automobile merely sucks [oxygen] from the air. But the rocket, designed to operate in space, must carry its own supply.
>
> —Willard Wilks, *The New Wilderness*[1]

The analogy is fairly simple. Getting into space involves understanding the basic physics of propulsion and mastering a specific type of mechanical engineering. Although a self-taught Russian mathematician developed the "rocket equation" in the late 1800s and a lone American physicist first demonstrated the "piston" technology for a liquid-fuel rocket in the mid-1920s, early civilian efforts still lacked adequate financial support to reach space.

If it were not for the fact that rockets can be used as ballistic missiles, space exploration might still be a dream. But the goal of Nazi Germany on the eve of World War II to bomb European cities that were more than a hundred miles away mustered the massive funding necessary to build the world's first rocket to approach the edges of Earth's atmosphere. It soon became the deadly V-2 missile.[2] After the war, both the

United States and the Soviet Union scooped up German military scientists, blueprints, and hardware to jump-start their missile programs and also bring space activity within their respective reaches. Their vast resources gradually led to the development of much-longer-range ballistic missiles designed to carry highly destructive nuclear weapons over intercontinental distances. These delivery systems offered lower vulnerability than bombers, promised to save the lives of pilots, and could transport weapons across the globe at tremendous speed, making attacks possible in less than an hour. Long-range ballistic missiles also made excellent space rockets. But the uncomfortable fact for scientists is that early space exploration emerged largely as a spin-off of these military programs. Spaceflight probably would have been accomplished before now by some country's scientists even in the absence of military incentives and large-scale government funding. But it would certainly have occurred much later than 1957 and with far fewer accomplishments to date. As the historian Walter McDougall argues in his Pulitzer Prize–winning book on the space age, "In these years the fundamental relationship between the government and new technology changed as never before in history."[3]

These points highlight an essential fact about space technology: its dual-use nature.[4] A space booster can launch an intercontinental missile or a civilian scientific probe to a distant planet. Similarly, communications satellites may broadcast a movie, transfer financial data for businesses, or transmit military orders (such as to an unmanned drone). An imaging satellite can survey Earth to monitor deforestation, facilitate city planning, or check on the progress of agricultural crops. But it can also allow military observers to track enemy troop concentrations, weapons deployments, and the success of ordnance already delivered against targets. The real questions are often not technical ones about a spacecraft's capabilities, but instead political and practical ones: who *controls* the spacecraft and to what *specific use* is it being put? Today, old lines separating military and civilian space programs are becoming increasingly blurred, as budget pressures and the growing sophistication of civilian technologies make it more efficient for military users to lease transponders on commercial satellites to carry military transmissions on an as-needed basis than to operate large military constellations for these purposes. The result is that everything from soldiers' e-mail messages home to information on enemy forces to civilian television and radio broadcasts may be on a single satellite.

In order to discuss space activity intelligently, we need first to understand the technologies required for spaceflight and the physics that affect it. Once basic orbital mechanics and rocket technology had been understood and mastered in the 1950s, a new set of challenges arose in developing equipment that would enable living beings (dogs, apes, and humans) to survive in the harsh environment of space. Given the risks involved, only three countries have thus far launched humans into space, with the Soviets getting there first and the United States launching the majority of astronauts who have reached space to date. China is the newest member of this small group and one with ambitious plans. Finally, we examine military space technologies and applications, which drove much of the U.S.-Soviet "space race" and today continue to motivate many national space efforts. Consistent with the notion of dual use, not all of these technologies are weapons. In fact, few are. The main benefit of space for national militaries has been and remains *information*. A key takeaway from this brief history is that while the first space powers had to invent all of these technologies, many of them can be purchased today, and that availability has accelerated the growth of spacefaring countries.

A BRIEF HISTORY OF SPACE SCIENCE AND TECHNOLOGY

Astronomy

In the past several hundred years—a mere blip in human history—scientists have gained a remarkable amount of information about Earth and its relationship to the rest of the universe. Though we won't delve deeply into a long history that has already been well covered by others, it is important to run through a quick review of these still fairly recent (and radical) changes in human understanding.[5]

About five hundred years ago, after spending millennia viewing Earth as the center of all creation, philosophers and astronomers began to undermine long-held beliefs and religious doctrines regarding the planets. While the Greek Aristarchus of Samos had conceived of a Sun-centric solar system in the third century b.c., the concept failed to take root and was forgotten. But by the mid-1500s a wider scientific community and the first reasonably powerful telescopes helped convince others that the similar observations of the Polish Catholic cleric Nicolaus Copernicus were true: the Sun (not Earth) must be at the center of our planetary system. Copernicus argued that the idea of larger celestial bodies racing around a stationary Earth made little sense, and that furthermore Earth must be moving, as well as rotating on its axis relative to the Sun to provide periods of day and night. Knowledge about Copernicus's ideas began to spread throughout Europe's budding scientific community.

By the 1590s, the German mathematician Johannes Kepler used observations by the Danish astronomer Tycho Brahe to prove further that because of the differential effects of gravity in relation to distance, the planets moved in elliptical orbits around the Sun rather than in circles. The Italian physicist, mathematician, and astronomer Galileo Galilei built on this knowledge to prove the rotation of the planets, while identifying through telescopic observation a range of celestial bodies (such as moons) never seen before in space, thus confirming earlier ideas about the distance of the stars. While his radical ideas ran afoul of the Catholic Church, forcing Galileo to live under house arrest, the truth could not be held back any longer. During the 1660s to the 1680s, the British philosopher and mathematician Isaac Newton developed new understandings of gravity and highly accurate laws of motion that created unprecedented levels of predictability regarding the celestial bodies and their relative movement with respect to Earth. With the cosmos largely in place, it now fell to engineers to get us there.

Launch Vehicles

Various peoples in Asia had developed simple bamboo rockets powered by gunpowder and other incendiaries by the 1200s.[6] But these weapons lacked range and accuracy and could not be controlled once launched. The British military leader William Congreve updated certain Indian designs that had been employed against his troops in the late 1700s by using metal tubes and more standardized production. Such rockets figured prominently—if ultimately unsuccessfully—in the famous British attack on Baltimore in September 1814. It was these rockets' "red glare" that eyewitness Francis Scott Key memorialized in "The Star-Spangled Banner."

But the next leap for rocket technology required a new conceptual foundation. A deaf Russian high school teacher, Konstantin Tsiolkovsky, became the unlikely father of this revolution in the 1880s by coming up with the "rocket equation," which described the thrust required for a rocket to leave Earth's atmosphere.[7] Tsiolkovsky was not an engineer, however, and while he understood the benefits of using supercooled liquid hydrogen and oxygen fuels to accomplish this task, his own limited financial resources and lack of the requisite tools and skills did not allow him to attempt building such a contraption.

Enter U.S. physicist Robert Goddard. Working at Clark University in Massachusetts in the 1920s, Goddard used his knowledge of both physics and engineering to build and launch the world's first liquid-fuel rocket in 1926.[8] He developed an odd A-shaped rocket powered by gasoline and liquid oxygen linked by metal tubing to create a controllable liquid-fuel engine. Ironically, despite his accomplishments, Goddard—like many of his predecessors—faced ridicule in the American press for even proposing spaceflight at all. Poorly versed but influential critics in the media at the time rejected the whole idea of propulsion in the vacuum of space by arguing that a rocket would have nothing to "push against" and would therefore quickly stop moving (neglecting the idea that it might push against itself). In the short term, Goddard was unable to prove them wrong, as his rockets still lacked the thrust needed to reach space. But, in 1936, German engineers Wernher von Braun and Walter Thiel used funding from the Nazi military to scale up Goddard's conceptual breakthrough and create the first rockets to pass the edges of the atmosphere. Their eventual A-4 rocket represented an invulnerable missile capable of bombing Western European cities from a distance of up to 200 miles, although their accuracy remained quite poor. Hitler used more than 2,600 of the renamed V-2 (Vengeance) rockets against French, Belgian, Dutch, and British cities during World War II, killing more than five thousand people (almost all civilians).[9]

After World War II, a new and more powerful series of space boosters was developed from military-purpose rockets like the V-2. Now working in the United States for the U.S. military, von Braun developed the Redstone intermediate-range missile in the early 1950s and the larger Jupiter C rocket to test reentry vehicles for planned nuclear warheads, which would pass through space. Although the Jupiter C had the capability to launch a satellite, the U.S. Army had no authorization to do so and thus pursued only testing of weapons delivery systems.[10] The Jupiter used a mixture of hyper-cooled (cryogenic) propellant in its first stage with solid-fuel upper stages.

Rockets that travel into space are normally configured into stacked sections called stages. The first stage, responsible for lifting the whole rocket and its fuel load, is the largest and requires the most thrust. After around two to three minutes, its job is done and—to remove unneeded weight—it separates from the rocket, falling back to Earth (or into the ocean).[11] The second-stage engine then ignites and carries the rocket closer to or into space. The final stages are normally for releasing payloads or positioning them in the proper orbit. The critical actions required to put a satellite into space usually take only six to eight minutes, although putting it into its target orbit may require an additional few hours or more, depending on its inclination and altitude.

Meanwhile, the Soviet Union undertook similar missile research. In 1948, a team of engineers under chief designer Sergei Korolev conducted the first successful Soviet test of an intermediate-range missile with a range of 550 miles, although this information remained a secret of the Stalinist regime.[12] The program continued at an urgent pace in order to counter the U.S. advantage in bomber forces. Korolev's team began to scale up the capability of successive rocket engines to achieve a range that could reach the United States. Finally, in August 1957, the Soviet Union successfully tested its R-7 missile (with a cluster of four first-stage engines) for use as a long-range delivery system. Soviet leader

Nikita Khrushchev gave chief designer Sergei Korolev permission to attempt the world's first satellite launch using the R-7. The four-engine, two-stage rocket with conventional kerosene fuel and a cryogenic oxidizer launched the simple *Sputnik 1* broadcasting satellite into orbit in October 1957, shocking the rest of the world. The Soviets quickly followed by launching the dog Laika aboard *Sputnik 2*.

In the United States, congressional opponents of President Eisenhower sharply criticized the administration for allowing this apparent technological gap to emerge, putting at risk both U.S. security and reputation.[13] The United States had planned to launch a civilian satellite aboard a modified Navy-derived Vanguard rocket, but had failed to provide the project with adequate funding or technical support. Its first attempt, in December 1957, failed miserably, with an embarrassing explosion on the pad. This situation forced the Eisenhower administration to turn reluctantly to former Nazi scientist von Braun. Using a modified Jupiter missile (or four-stage Juno 1 rocket), von Braun's team successfully launched the first U.S. satellite (*Explorer 1*) on January 31, 1958.

As the Cold War rivals raced to launch ever larger and more complex payloads, the Soviets undertook the first human spaceflight (Yuri Gagarin in April 1961), and the competition soon escalated to multiperson spacecraft and planned Moon landings. But for these missions, they needed more powerful boosters. Because of a variety of technical and organizational problems, the Soviets failed in their efforts to develop the multi-engine, three-stage liquid-fuel N-1 lunar rocket.[14] The United States succeeded, however, in developing the massive three-stage, liquid-fuel Saturn V for its lunar missions, including its December 1968 orbital flight around the Moon and culminating in NASA's historic Moon landing in July 1969. Several additional lunar missions followed. The Saturn V could lift 300,000 pounds (or the equivalent of about 200 satellites) into low-Earth orbit, making it the most powerful launcher ever built (table 1.1). The Saturn V had to carry the fuel required to get the *Apollo* spacecraft and the lunar lander to the Moon, with enough spare propulsion to return part of the lander up to the orbiting crew capsule and then to ferry the astronauts safely back to Earth.

Soviet rockets included the workhorse Soyuz (1963 to the present), larger Proton (1965 to the present), and heavy-lift Energiya booster (1987–88), built to launch the short-lived *Buran* spaceplane. The Russians have continued to rely on variants of these liquid-fuel rockets. On the U.S. side, the space shuttle's launch system combined a massive liquid-fuel core with two solid-fuel rocket boosters to provide enough thrust to put a reusable spaceplane into orbit. The recent U.S. Atlas V rocket—often used to launch military satellites—employs a liquid-fuel core and several smaller strap-on solid-fuel boosters, as does the U.S. Delta rocket. By contrast, U.S. Minuteman ballistic missiles use solid fuel only.

Other leading spacefaring countries also use a combination of liquid- and solid-fuel rockets, depending on the purpose.[15] Typically, military missile programs prefer solid-fuel boosters—dry propellants cast with an oxidizing agent into molded compounds—because solid propellant can be more easily stored in a ready position for launch for long periods of time.[16] Civil space programs and military satellite programs normally use liquid-fuel systems (sometimes supported by specific solid-fuel elements) because of their greater controllability in terms of throttling and comparatively greater thrust-to-volume ratio. But the use of liquid fuel typically requires the "tanking" of the rocket before flight and, frequently, the use of a cryogenic oxidizer (usually liquid oxygen), which must be supercooled and therefore cannot be stored for long once the rocket is fueled. An exception is the use of storable hypergolic fuels, such as those used by the early U.S. Titan missile and the current Russian Proton booster, which consist of two chemicals that ignite when they are mixed. But some of these fuels are highly corrosive, which creates additional problems. Nevertheless, the use of solid propellants in ballistic missiles and liquid propellants in space-launch rockets is the general norm.

This liquid-solid fuel distinction came into play during the 1990s in international negotiations on application of the Missile Technology Control Regime (MTCR), an initiative started by the United States in 1987 to prevent the spread of missile delivery systems. In order to persuade more countries to forgo solid-fuel military missiles, the MTCR agreed to allow new members particularly to continue with

TABLE 1.1: Selected Cold War Space Launch Vehicles

Vehicle (Country)	First launch	Height (ft.)	Thrust (lbs.)	Fuel	Payload capacity (lbs.)
R-7 (USSR)	1957	108	877,000	Liquid	3,300
Juno 1/Jupiter C (U.S.)	1958	71	111,000	Liquid and solid	<100
Soyuz (USSR)	1963	147	1.2 million	Liquid	15,000
Titan II (U.S.)	1965	90	530,000	Liquid	4,200
Diamant A (France)	1965	62	111,000	Solid and liquid	<200
Saturn V (U.S.)	1967	334	9 million	Liquid	300,000
Proton (USSR)	1967	174	1.5 million	Liquid	45,000
Long March 1D (China)	1970	92	476,000	Liquid (solid third stage)	1,650
Lambda 4S (Japan)	1970	54	168,000	Solid	<100
Satellite Launch Vehicle 1 (India)	1980	74	110,000	Solid	<100
Space Shuttle (U.S.)	1981	184	6.7 million	Solid and liquid	65,000
Energiya (USSR)	1987	197	8.9 million	Liquid	300,000

SOURCES: James Clay Moltz, "Managing International Rivalry on High Technology Frontiers: U.S.-Soviet Competition and Cooperation in Space" (Ph.D. diss., University of California, Berkeley, 1989), table 1, "Selected U.S. and Soviet Space Launch Vehicles," 55; Kenneth Gatland, "The World's Space Launch Vehicles," in *The Illustrated Encyclopedia of Space Technology*, 2nd ed. (New York: Orion Books, 1989), inside cover and 304–6.

liquid-fuel space-launch vehicles for peaceful purposes.[17] Of course, this solution is imperfect. Liquid-fuel rockets can indeed be used for weapons delivery in a pinch, as with early-generation U.S. and Soviet long-range missiles. But they are vulnerable to attack while being fueled, are more difficult to transport, typically rely on fixed launch pads, and can remain ready for launch for only a certain period of time before they have to either be used or undergo a delicate defueling process (delicate because the liquid fuel can explode). In October 1960, an impatient

Soviet general under pressure from political authorities ordered his scientists and engineers to investigate why the new liquid-fuel heavy-lift R-16 rocket had not launched, disregarding their admonitions that the fuel needed to be drained out first.[18] Soon after they left their protective bunkers, the liquid fuel in the booster's second stage ignited, setting off an enormous fireball that killed 126 scientists, engineers, and soldiers. But solid-fuel propellant can also be dangerous. A similar fatal incident occurred in Brazil in 2003, on the eve of an attempt to carry out the country's first satellite launch. A solid-fuel booster accidentally ignited while undergoing final tests,[19] killing 21 of Brazil's top space scientists and engineers and setting the program back more than a decade.

SATELLITES AND ORBITS

Beyond human spaceflight, the main purpose of developing space boosters is to put satellites into orbit to either gather information or broadcast it.[20] There are currently about 1,050 operational satellites in various orbits ranging from around 200 to 22,300 miles above Earth's surface.[21] Those satellites that gather information can be compared to big "cameras," but with sensors much more sophisticated than those on your cell phone. They seek a variety of types of information about features on Earth's surface generated by different forms of electromagnetic radiation. These include visible light, infrared emissions, and radio waves. Some of these satellites also send out radar signals to Earth and gauge features of the returned information to provide "active" imaging through darkness and adverse weather conditions. On the broadcasting and communications side, satellites act essentially like television or cell phone towers on Earth, sending radio waves that transmit information to receiving stations on Earth (such as handheld devices or rooftop dishes). This information could be financial data, a weather forecast, a phone call, a radio broadcast, a movie, precise locational information, or an Internet connection. The bulk of today's satellites are powered by on-board solar cells, although some satellites operate on conventional batteries or, on rare occasions, on heat generated by the decay of radioactive materials (such as plutonium, uranium, or cesium). The Soviet Union, whose solar-power-generation capability was initially poor, was the only country to make use of active nuclear-fission reactors (fueled by highly enriched uranium) to power radar satellites tracking U.S. naval operations in the 1970s and 1980s. More frequently, deep-space programs have used the passive decay of radioactive materials (in so-called radioisotope thermal generators) to produce needed electricity far away from the Sun's energy.

Satellites travel in a variety of orbits that relate to their function.[22] Orbits can be divided by altitude above Earth and by inclination (or the angle of the orbit in relation to the equator). The easiest region to reach is low-Earth orbit (LEO), which begins just above the end of Earth's atmosphere at around 60 miles up and ranges to an altitude of about 1,200 miles (figure 1.1). Being closer to Earth's surface is beneficial for observation and for communications with small mobile devices that lack large antennas. Spacecraft circling Earth today in LEO include military reconnaissance spacecraft, civilian remote-sensing satellites, and mobile telecommunications systems (such as the satellites in the Iridium constellation). Half of all satellites are in LEO orbits.[23] The *International Space Station* (*ISS*) also orbits in LEO. Precision timing and navigation constellations are typically located in medium-Earth orbit (MEO), which spans from approximately 1,200 to 22,300 miles up, although almost all such satellites are located in the middle of this range. Here they can survey a larger swath of territory and triangulate with other satellites (using highly precise atomic clocks) to provide specific locational data, while still being able to communicate effectively with receivers on Earth. Currently, only seven percent of all satellites are in MEO orbits.[24] U.S. Global Positioning System (GPS) satellites are located in MEO orbits at approximately 12,000 miles up. The highest orbital band in frequent use is geostationary orbit (GEO) at 22,300 miles up, where satellites travel at the same speed as Earth's rotation and therefore appear to be motionless, allowing them to "stare" continuously at large, continent-sized areas on the ground. This makes GEO orbit ideal for satellites engaged in missile early-warning, nuclear testing detection, electronic intelligence, commercial communications, and direct broadcasting (such as direct television and radio services). Nearly half of all satellites are located here.[25]

In terms of inclinations, satellite orbits are categorized according to their relation to Earth's equator (figure 1.2). For example, a GEO satellite moving in synchrony with Earth's rotation above a specific point above the equator is said to be in a zero inclination orbit. By contrast, an observation satellite that flies over Earth's poles is in 90-degree inclined orbit relative to the equator. Such satellites have the advantage of passing over every point on the globe during a single day. Smaller inclinations (less than 90 degrees and greater than zero) may be useful for maximizing coverage over specific regions or hemispheres. Those satellites that orbit the same region during daylight hours are in so-called sun synchronous orbits, which are especially useful for imaging satellites. Other orbits include highly elliptical paths that allow satellites to linger for longer periods above a certain region and then scoop quickly around the other side of Earth at very low altitude. The Soviet Union, which had difficulty covering its polar regions from above the equator in a GEO orbit, used specifically inclined, *molniya* (or lightning) elliptical orbits to handle communications across its vast territory. The high latitude of Russian launch sites was another constraining factor, which made highly inclined orbits easier to reach.

Earth satellites require a certain rate of velocity to achieve orbit and remain there. That's why you cannot shoot a bullet into space—it falls back to Earth (somewhere!). Objects that lack this minimum velocity (a better example is an intercontinental ballistic missile) will travel in an arc through space and reenter the atmosphere somewhere across the globe. A satellite in low-Earth orbit must travel even faster—around 18,000 miles per hour--in order to surpass forces of atmospheric drag and gravity that would cause it to deorbit.

Changing orbital planes is a difficult task that requires the use of propellant. As the physicists David Wright, Laura Grego, and Lisbeth Gronlund explain, "Because the orbital speed of satellites is so large, the velocity changes required for maneuvering may also be large, requiring the thrusters to use large amounts of propellant."[26] Since fuel is heavy, satellite operators seek to minimize the movement of a spacecraft over

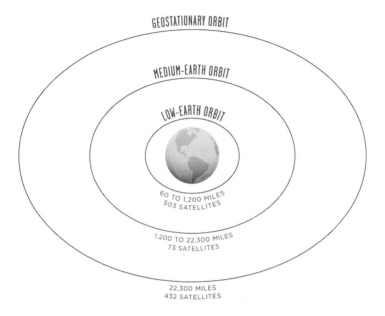

GEOSTATIONARY ORBIT

MEDIUM-EARTH ORBIT

LOW-EARTH ORBIT

60 TO 1,200 MILES
503 SATELLITES

1,200 TO 22,300 MILES
73 SATELLITES

22,300 MILES
432 SATELLITES

FIGURE 1.1
Data: Union of Concerned Scientists

the course of its lifetime and to use the most fuel-efficient means possible. By using Earth's rotation to "help" with the process (through a so-called Hohmann transfer orbit), a move can be completed with less fuel than attempting to travel directly into the new orbit. Typically, military reconnaissance satellites have the highest fuel demands because they frequently need to be tasked with high-priority assignments that may require them to pass over specific locations in a short period of time. For other satellites, besides the initial burning of propellant required to deliver them into their proper orbits, most of their fuel is used for station-keeping, or ensuring that they remain where they need to be, given the slight drift that occurs in all orbits because of atmospheric drag and other factors. According to U.S. licensing rules and the 2007 UN debris guidelines, satellite operators are also instructed to save a certain amount of fuel for end-of-operation maneuvers to either push their

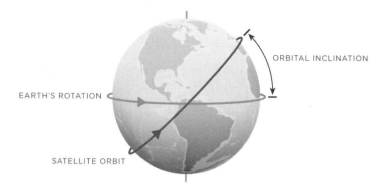

FIGURE 1.2: Orbital Inclination
Source: NASA

satellite up into a super-GEO parking orbit or, for a LEO and MEO satellite, to push it down to burn up in reentry into Earth's atmosphere. These critical operations help limit the amount of orbital debris and reduce the chances for collisions with other satellites.

ORBITAL DEBRIS

The lack of atmosphere and the weakening effects of gravity as objects move farther from Earth cause high-velocity projectiles like satellites to remain in orbit long after they have ceased to be useful. While orbital objects below about 500 miles will be affected by atmospheric drag and will deorbit within several decades (and more quickly at lower altitudes), satellites above this altitude will remain in orbit for centuries or longer. Unfortunately, that means that lots of orbital debris—both old satellites and fragments of space junk from stage separations, inadvertent loss of spacecraft integrity, fuel-tank explosions, or collisions—will remain in orbit for many years, putting at risk other spacecraft that are still functioning.[27] For example, the nearly thirty Soviet nuclear-powered radar ocean reconnaissance satellites from the 1970s and 1980s remain at orbits of about 1,000 miles, even though their reactors have long since ceased functioning.[28] But their leaking sodium coolant and their onboard highly enriched uranium cores pose both short- and long-term risks. Other objects as small as bolts and even paint flecks represent speeding bullets that could damage or destroy any spacecraft in their path, given their tremendous velocity. Keeping track of what is in space and what risks are involved as thousands of large and small objects hurtle around Earth are two of the more significant "security" challenges of the twenty-first century, particularly in heavily trafficked areas like LEO. The U.S. Joint Space Operations Center (JSpOC) tracks operational satellites, dead satellites, and debris using mainly ground-based radar. It also attempts to prevent these various objects from colliding with one another by getting operators to move their spacecraft using minor propellant burns. This requires timely notification and cooperation, which has not always been the stock-in-trade of military space operators. However, after the collision of a U.S. commercial satellite with a dead Russian spacecraft in 2009 that produced thousands of large pieces of debris, the JSpOC established new rules for sharing information and is working through U.S. government channels to make sure the changes reach operators in a timely manner. But some debris is too small to track (if it is less than about half an inch in diameter).[29] Unless a satellite is armored against such minor collisions, tiny pieces of debris can still cause significant damage. Satellite operators have to judge the likelihood and seriousness of a collision themselves and decide whether or not to use their precious fuel supplies to try to avoid it. But the debris problem is not going to go away; in fact, it may have reached a critical tipping point, due to the possibility of cascades of micro-debris being created by increased collisions among already extant orbital debris particles. This is the so-called Kessler syndrome, predicted by a NASA scientist in 1978. Many experts believe it is now happening.[30]

HUMAN SPACEFLIGHT

Although suborbital flights with animals into space began in the 1950s in both the United States and the Soviet Union, the first orbital flight of an animal took place in November 1957 with the dog Laika in a one-way mission aboard *Sputnik 2*. Problems with the cooling system[31] and other concerns kept Soviet engineers busy until April 1961, when Yuri Gagarin became the first human being to orbit Earth and return safely to the ground aboard *Vostok 1*. Since that time, nearly 550 people have traveled into space aboard Soviet/Russian, U.S., and, most recently, Chinese boosters.[32] The next decade will see an unprecedented increase in both the number of countries and the number of private companies taking people into space.

Placing human beings into space, obviously, involves additional requirements for spacecraft. People need to be protected from excessive gravitational changes (or "g forces"), be kept within a certain temperature range, and have access to food, oxygen, and (for flights longer than a few hours) some means of handling waste. These functions add cost to manned missions, compared to robotic spaceflight, and they also involve greater risks. As the United States found in 1967 in its fatal ground test of the *Apollo 1* spacecraft, using an all-oxygen environment for astronaut compartments was easier than carrying air with its inert components, but it resulted in a dangerous fire risk. Subsequent missions have used a less flammable mix of oxygen and nitrogen.[33] Soviet cosmonauts died in reentry and landing accidents during the Cold War, and the United States lost two of its shuttles and their crews (*Challenger* in 1986 and *Columbia* in 2003) as the result of engineering problems associated with their solid-fuel propulsion system and external heat-shielding tiles, respectively.[34] Any country that undertakes human spaceflight accepts risks and has to operate at a higher level of safety than those with only satellite or robotic missions.

The Soviet Union launched the world's first space station (*Salyut 1*) in 1971, but suffered three fatalities associated with the reentry vehicle. Following three successful U.S. missions aboard the huge *Skylab* space station (built from the upper stage of a Saturn V rocket) from 1973 to 1974, the Soviet Union conducted successful missions on a series of new *Salyut* stations, as well as the larger, module-design *Mir* station from 1986 to 2001.[35] The U.S. space shuttle carried more than 350 people into space during its twenty-year operational period from 1981 to 2011, an impressive 70 percent share of the total astronaut population to that date.[36] Its large payload bay also provided the opportunity to place satellites into orbit, conduct experiments, and deliver the key building blocks for the *ISS*.

The *ISS* is a remarkable achievement of international space cooperation. Originally begun as the U.S. *Freedom* station, then the multinational *Alpha*, it eventually became the *ISS* after adding Russia to the team of nations (the United States, Canada, the European Space Agency, and Japan) that had been formed in the mid-1980s to build the multi-module spacecraft. Assembled in parts from 1998 to 2011, the station now represents the most expensive international engineering project in history. The station includes the original *Zarya*, plus the *Zvezda*, *Pirs*, *Poisk*, and *Rassvyet* modules built by Russia; the U.S. *Unity*, *Destiny*, *Harmony*, and *Tranquility* sections; the European *Columbus* module; the Japanese *Kibo* module; and the Canadian manipulator arms. Together, these components weigh in at a massive one million pounds.[37] The station marks a sharp departure from the highly nationalistic human spaceflight programs that characterized the Cold War. However, such emerging spacefaring nations as India and China are not yet members of the *ISS*. After failing to receive an invitation to join by the members, China has developed plans for its own 60-ton space station by 2020. But it may eventually participate in the *ISS* as well, depending on future space politics, although not likely as a full voting partner. The station is scheduled to remain in operation until at least 2020 and perhaps longer.

The newest evolution in human spaceflight is the recent drive for private, commercial space services. Although the Soviet Union first marketed spare slots on its *Mir* space station in 1990 for about $20 million, Russia has continued to offer seats on its spacecraft going to the *ISS* for fees of over $60 million (including the cost of extensive training).[38] But much lower-cost access may now be emerging. Two types of services are on tap: (1) flights aboard commercial boosters (such as the U.S. Falcon 9) to the *ISS* or new private space hotels at a cost that will be perhaps one-twentieth that of current Russian fees; and (2) suborbital flights offering several minutes in space before returning to Earth for around $250,000 (and eventually less). Such services will not bring human spaceflight to the "masses" in the coming decade, but these new options and their considerably lower cost and time commitments (as short as three days for suborbital flights, including training) will increase the numbers of people who can enter space to as many as a thousand per year (with further growth over time). The effect of such an influx of ordinary (if wealthy) citizens into space is unclear. One impact will certainly be a higher priority than ever before on the mitigation of orbital debris and the testing of any debris-producing weapons in low-Earth orbit. Such hazards will put many more people and businesses at risk, placing new restrictions on both civil and military activities, at least in the lower reaches of space.

SPACE WEAPONS AND DEFENSES

Early in the space age, the United States and the Soviet Union began to experiment with military uses of space that went beyond passive systems and information-oriented support functions. As early as World War II, in fact, Nazi German engineers had developed plans for a military space bomber aimed at exo-atmospheric bombing runs on the United States.[39] But the war ended without any such technical developments. Still, the two Cold War superpowers realized from the experience of the V-2s that larger rockets could be used for the long-range delivery of military payloads, particularly nuclear weapons. Such missiles—whether launched from ground, sea, or air—could also be used to attack objects in space. With a nuclear weapon on board, the missile need not even hit its target, but could create crippling damage from afar—not through kinetic destruction, but by the disabling effects of electromagnetic pulse (EMP) radiation on a satellite's electronics. Such weapons entered the arsenals of both superpowers in the late 1950s, and the two sides together tested nine nuclear weapons in space from 1958 to 1962 to learn about the effects of EMP and to test the anti-ballistic missile capabilities of ground-based, nuclear-tipped interceptors.[40] The 1.4-megaton U.S. Starfish Prime test launched from Johnston Island in the Pacific in July 1962 created such large EMP emissions that it eventually disabled seven satellites in LEO, including U.S. military satellites, the first U.S. civilian communications satellite, and British and Soviet spacecraft. The incompatibility of space nuclear tests both with human spaceflight (because of health concerns) and with unmanned military and commercial satellites (because of problems posed for their electronics) caused both sides to step back from the brink of ruining orbital space. By banning further nuclear tests in space under the 1963 Partial Test Ban Treaty, the leading spacefaring nations accepted mutual restraint and averted disaster.

But such limited cooperation did not prevent the two sides from continuing to develop and deploy anti-satellite systems. In the U.S. case, this work proceeded through proximity tests of rockets against space-based targets, with the assumption that a nuclear warhead would destroy satellites within the line of sight of any such detonation, as well as eventually disable any other satellite passing through its charged EMP wake over the next months and years. The United States had two nuclear-tipped anti-satellite (ASAT) programs during the 1960s—code-named Project 505 and Project 437—but it eventually decided that the isolated launch site on Johnston Island in the Pacific was vulnerable to attack and that in any case the actual use of the weapons would cause more damage to other U.S. satellites than could justify the destruction of any specific Soviet satellite.[41] The Soviet Union developed a conventionally armed ASAT weapon in tests from 1968 to 1982.[42] The satellite would be launched into orbit and then gradually move into the orbit of its target satellite over several hours, finally launching an explosive projectile when it got nearby. The Russians eventually mothballed this weapon in the early 1990s due to its cost and their changing relationship with the United States. During the Reagan administration, the United States also tested a kinetic ASAT system that used a miniature homing vehicle launched into space from an F-15 aircraft designed to crash directly into its target. In its one active test in 1985, the U.S. military hit the target satellite, destroying it and creating hundreds of pieces of hazardous orbital debris.[43] That was the last such kinetic ASAT test until 2007, when China broke an informal international moratorium and destroyed one of its own satellites using a homing device aboard a ground-launched mobile missile. U.S. radar systems identified more than 3,000 pieces of orbital debris from this collision, most of which will remain in orbit for about fifty years.[44] This high-velocity debris will continue to pose a hazard to all spacecraft in the lower reaches of space until the particles deorbit. While a number of scientists and policy experts have called for a ban on such kinetic weapon tests, no agreement has yet been reached. This situation raises the risk of similar tests in the future unless countries begin to change their behavior.

Over the years, there have been many proposals for deploying weapons in space for both defensive and offensive purposes. These ideas have ranged from missile bases on the Moon in the late 1950s to orbital battle stations to be manned by uniformed military personnel in the 1960s to manned and unmanned military spaceplanes later on. By the 1980s, U.S. defense officials outlined ambitious proposals for hundreds of space-based lasers and thousands of orbital interceptors (called Brilliant Pebbles) as part of the Reagan administration's Strategic Defense Initiative. During the George W. Bush administration, some senior U.S. military officials discussed using "Rods from God" (orbital bundles of tungsten rods) for possible attacks on terrorists or deeply buried targets located within rogue states.[45] Due to considerations of cost, launch requirements, technological immaturity (or impossibility), and foreign military reactions, none of these systems has ever gotten off the ground. Yet a number of these weapons continue to find support in some quarters in the United States and among some military leaders in other spacefaring nations as well. Thus far, the consensus among military and political officials across the globe is that it is not a good idea to push for these systems, as their effects on the heavily trafficked LEO region could be quite harmful, especially if they are eventually tested and deployed in large numbers by *multiple* countries. Yet, no international agreements exist that prevent countries from launching or testing any of these weapons, with the exception of nuclear texts or the space-basing of any weapons of mass destruction.

Other types of passive defense systems have also been proposed for space. These include additional fuel storage to improve spacecraft maneuverability, satellite decoys to confuse potential adversaries and raise attack requirements, ground-based spares to allow for quick replenishment of assets attacked in space, metal chaff or low-visibility paint or construction materials to confuse radar and infrared seekers, or shutters to keep out damaging laser interference.[46] A number of these systems are being pursued by countries today because of their lower cost, less threatening profile, and likely greater effectiveness when compared to kinetic weapons. Nevertheless, given that most spacecraft are shiny objects orbiting around Earth against a black and mostly empty background, certain inherent vulnerabilities will always be a part of operating in space. The question for the future is how best to address these risks: through military technology, diplomacy, or perhaps some other means?

CONCLUSION

Despite all of the remarkable developments in space activity since 1957, human beings on Earth are still in the early stages of becoming a spacefaring civilization. Fewer than seven people are currently able to live permanently in space (aboard the *ISS*). No one is capable of living on any celestial bodies, and the concept of people living independently from Earth is certainly many decades away. But what began as almost a purely competitive process during the hostile U.S.-Soviet Cold War has gradually evolved into a mixed environment of competition and cooperation. Perhaps the most obvious sign of a shift since the Cold War is the *ISS*, which has essentially merged the world's two largest human spaceflight programs (those of the United States and Russia). Yet space competition in the military, commercial, and civil space fields has not disappeared. A range of additional countries, companies, universities, and even private citizens are now entering the space field, changing the face of space actors and activities.

As small "cubesats" (a mere 10 centimeters on each side) and other low-cost satellites become more widely available in the coming years, the financial threshold to entering space will drop further. But independent launch access is likely to remain a more limited field, comprising in 2013 only ten countries (the United States, Russia, China, France, Japan, India, Israel, Iran, North Korea, and South Korea). Yet such nations as Brazil, Indonesia, and Pakistan have plans to join this club, perhaps soon. Only Russia and China currently offer orbital human spaceflight access, but the United States will soon reenter this field, and may be joined by India, the European Space Agency (ESA), Japan, and others in the coming decade.

Overall, the most obvious trend is the continued spread of space technology to new and varied actors. The challenge this poses to the international community has to do with how new actors will behave in space and whether existing mechanisms for preventing space conflict will be adequate to the task. As seen in Asia, Latin America, Africa, and other developing regions, space technology is viewed as a key asset for a modern nation to acquire, particularly because of its perceived economic importance. Like nuclear energy technology, it has frequently become a chit in regional competitions for prestige and military power, which raises the potential for conflict.

But most space actors continue to be motivated by the desire for safe access to space for functional purposes—to acquire information and to conduct activities. Like drivers on a highway, they want to get where they are going safely. These incentives for peaceful activity and non-interference with their spacecraft may dampen risks of conflict in space. But even small numbers of hostile actors have the potential to cause great damage to space because of the shared nature of this environment and the inability of space to cleanse itself of orbital debris, except over long periods of time. For these reasons, "getting along" in space will be an increasing imperative as orbits becomes increasingly crowded. Like the opening of railroads, automobile highways, and jet airways over the past two hundred years, the increase in space traffic will require new mechanisms to create safety standards and enforcement procedures.[47] The great challenge in space is that national jurisdiction does not exist, except in the licensing process for individual space launches. Thus, innovative solutions will be needed at the international level.

In order to project where we might be headed in space governance, we need first to understand the mechanisms we have. Fortunately, there is a significant international framework for encouraging peaceful activities in space and discouraging harmful behavior. But these treaties, conventions, and guidelines have considerable gaps and have not kept up with the recent spread of space technology. Many of the existing agreements lack enforcement mechanisms, meaning that compliance is voluntary. This situation may not be a problem when there is strong self-interest at work to follow the rules of the road. But history suggests that in the absence of adequate methods for verification (think: police, speed guns, and traffic cameras) and enforcement (tickets), cheating could occur. It remains unclear who will or can play the role of a "space cop" or international police force, or whether instead such functions could be crowdsourced to scientists and the interested public, including amateur astronomers. Fortunately, radars based on the ground or in space can follow much of what is happening, and with much better effectiveness than in other environments. Cameras could also conceivably be put in space. Yet only a handful of countries currently have the tools to provide anything close to comprehensive coverage, and only the United States currently operates a constantly updated catalog of space objects. Russia, the European nations, and China have some capabilities. Interestingly, a consortium of commercial satellite companies is now beginning to create its own system for monitoring areas of space that affect their business interests and to sign up like-minded firms, and there are space-tracking websites maintained by scientists and devoted space watchers around the globe. But it remains to be seen whether these capabilities will evolve into an effective network for attribution of wrongdoing and the sanctioning of bad actors.

2

THE POLITICS OF THE SPACE AGE

The cold war was an unmitigated blessing for both sides' rocketeers . . .
because they fed off and were nourished by the competition. Rockets were
almost always developed for the wrong reason during the cold war. But they
were developed.

—William E. Burrows, *This New Ocean*[1]

The USA and USSR have gone further to achieve arms control in space
than in any other area.

—William H. Schauer, *The Politics of Space*[2]

Only a little more than a decade after the Wright brothers' first air-
plane flight in 1903, militaries were shooting at each other from rickety
wooden contraptions in the skies above Europe, killing one another
and, occasionally, people on the ground. The lethal power of aircraft
expanded exponentially during World War II, eventually leading to the
dropping of two atomic bombs that killed more than 100,000 Japanese
instantly. Despite many predictions of space war, however, such direct
conflict has not taken place yet in orbit, more than fifty-five years after
the first spaceflight. Whether this record can be extended and military
restraint preserved is a critical question that will shape the character of
future space activity.

The political history of Cold War space activity tells a puzzling
story: how is it that two hostile adversaries managed to avoid an
arms race in an area that both of them recognized as critical to their

national security? Military restraint was certainly neither side's first intention. Much of what we now take as conventional wisdom about space was actually learned only by trial and error. Fortunately, we got lucky and neither ruined space permanently nor killed one another. But this delicate balance between healthy competition and lethal conflict required constant attention to prevent matters from spiraling out of control. U.S.-Soviet treaties, international conventions, and regular communications helped prevent the burning hostility between the two superpowers from being acted out in space. But only just barely. How we managed may provide some useful lessons for the future.

After the breakup of the Soviet Union in late 1991, the United States emerged as a powerful hegemon in space, although one facing serious budget problems from its surge of weapons buying in the 1980s as it sought to muscle the Cold War to an end. The end of hostilities pushed the two sides to collaborate on building the *International Space Station*, effectively burying their conflict.

The current period of international relations in space began in 2003 with China's first human spaceflight. Beijing's subsequent emergence as a military space power has posed a potential challenge to the United States and has cemented a new era characterized by multiple significant space actors with a much wider range of capabilities than ever before. Economic globalization has internationalized the space marketplace, creating new economies of scale, spurring innovative services, and promoting the rapid spread of space technology, including in the military sector. But this process of transition and the changes in space dynamics—for better and for worse—have raised questions about the adequacy of old mechanisms for managing space competition in the twenty-first century.

A ROCKY START IN SPACE

Like Columbus's landing in the New World in 1492, the Soviet launch of *Sputnik* in 1957 opened a vast new frontier for international competition. Space, like any other frontier, started out as an ungoverned environment—one without posted signs or rules. Military missiles had led the way there, suggesting imminent conflict, but some people believed that the world's scientists might forge a path to new forms of international cooperation in space, as they were doing on the Antarctic continent at the very same time. These hopes stemmed from an ongoing global scientific exchange called the International Geophysical Year (IGY) and related efforts to prevent conflict in Antarctica by negotiating a treaty that set aside prior territorial claims and banned military activities.[3]

In January 1957, the United States had proposed that any developments in outer space "be devoted exclusively to peaceful and scientific purposes" and that any testing of space systems be placed "under international inspection and participation."[4] The Soviet Union, however, had no intentions of opening up its military missile program to prying Western eyes. Immediately after *Sputnik*'s launch, U.S. ambassador to the United Nations Henry Cabot Lodge, cognizant of the leading role of the Soviet military in its space activities, reiterated the U.S. proposal for the formation of an international body charged with inspecting satellites and limiting space activities to nonmilitary purposes. Moscow again demurred. A few months later, Soviet ambassador to the United States Andrei Gromyko proposed a ban on military activities in space linked to a prohibition of military bases on foreign soil. His proposal was clearly aimed at the North Atlantic Treaty Organization (NATO), through which the United States was deploying nuclear weapons in several European countries at the time. In the highly distrustful climate of the Cold War, this proposal went nowhere as well.

Soon the defense establishments of the two sides began to undertake weapons-related experiments as planners began to work on the assumption that war in space might be inevitable. Lieutenant General Donald L. Putt of the Air Force gave a speech in early 1958 in which he called for establishment of a U.S. missile base on the Moon to fire nuclear-tipped rockets at the Soviet Union in the event of war.[5] But scientists such as Cal Tech president Lee DuBridge fought back, countering that "if you did launch a bomb from the moon, the warhead would take five days to reach the earth. The war might be over by then."[6] General Putt's proposal quietly died.

In late 1958, the two superpowers grudgingly agreed to the formation of the United Nations Ad Hoc Committee on the Peaceful Uses of Outer Space. But when the organization attempted to meet, Soviet opposition shut it down, protesting that the group's overwhelmingly Western-leaning membership would skew voting. By then, both sides had already proceeded with military space programs, and the United States had conducted the first nuclear weapons tests in space, in the late summer of 1958.

No arms control treaties yet existed between the two superpowers, and expectations were low for any agreements, given the hostility of political relations between Moscow and Washington. The Eisenhower administration's decision in late 1958 to create a wholly civilian space agency—the National Aeronautics and Space Administration (NASA)—to compete with the Soviet Union's military space program marked the first major divergence in the approaches of the two sides. But it did not lead to cooperation. Instead, the Soviet Union made the most of its initial advantage in lift capacity to outpace the United States in its launch of a variety of payloads. Space became symbolic of the struggle between communism and democratic capitalism, making cooperation highly unlikely.

President John F. Kennedy's response to the Soviet launch of cosmonaut Yuri Gagarin in the spring of 1961 revealed a profound crisis of confidence in the United States. But Kennedy—encouraged by his pro-space vice president, Lyndon Johnson—took the bold and risky decision to pledge the United States to become the first country to send an astronaut to the Moon by the end of the 1960s, even before any American had entered space. Somewhat to his surprise, the U.S. Congress endorsed his plan and provided the funds needed to begin what became the Moon race. As the space historian John Logsdon has argued, Kennedy was "a very unlikely candidate to send Americans to the Moon," given his limited interest in space.[7] But he needed a victory over the Soviets to shore up his struggling administration after the failed Central Intelligence Agency–led disaster at the Bay of Pigs in Cuba in April 1961. Although Kennedy considered cooperating with Moscow in the space venture, Congress would have none of it. NASA's budget soared, and the United States put a major focus in its political rhetoric on the importance of the space program as a "test" of the United States. In the end, First Secretary Khrushchev in Moscow did not take the bait of Kennedy's offer, instead seeking to extend the apparent Soviet lead in space to his own advantage.[8]

The first devoted space resolution at the United Nations included elements from both U.S. and Soviet proposals.[9] Passed in the fall of 1961, UN General Assembly Resolution 1721 called upon countries to adopt the principles that international law (including the UN charter) applied to space and that outer space should not be subject to national appropriation, while inviting the UN space committee to study further other legal issues that might arise in space. In addition, the resolution exhorted all countries launching objects into space to register their activities with the United Nations and the space committee to create a public registry. These first steps were advisory only, given the resolution's lack of legal status and the absence of any obvious mechanisms to enforce compliance.

Behind the scenes, the U.S. and Soviet militaries raced to develop space defenses: nuclear-tipped anti-ballistic missile systems for use in space, kinetic anti-satellite weapons, intelligence-gathering satellites of various types (imaging and signals), and even to plan military space stations. While lagging in high-prestige civil space activities, the United States actually led the Soviet Union by several years in the secret operational development of a variety of military support technologies. Most of these activities, such as the Navy *Grab* (*Galactic Radiation and Background*) satellites and the Central Intelligence Agency (CIA)/Air Force *Corona* satellites, remained highly classified, hiding behind fake names created to put the public off the scent. The *Grab* spacecraft were developed by the Naval Research Laboratory and, beginning in 1960, collected information on Soviet air-defense radars for understanding Soviet military plans and targeting critical facilities in case of war.[10] The *Corona* flights were all launched under the Discoverer scientific program, a purported U.S. effort to conduct medical and biological experiments to lead the way for human spaceflight. Similarly, the Soviet Union conducted its secret military experiments under the generic Cosmos rubric, which covered everything from scientific probes to its spy satellites.

The United States sought to restore its image as a technology leader among its allies by offering the services of the first U.S. civilian communications satellite (*Telstar 1*) in 1962. Still, the political environment surrounding space activity remained tense. After Soviet nuclear tests in space in 1961 and the ramping up of military tensions, U.S. secretary of state Dean Rusk cautioned: "There is an increasing danger that space may become man's newest battlefield."[11] This situation raised the specter of a stark trade-off: either space was going to continue to be a weapons-testing venue or it would be developed for scientific, commercial, and passive military purposes. It could not be both.

LEARNING TO SHAKE HANDS WITH THE ENEMY

The decision to step back from the precipice of space conflict did not come immediately or easily. The loss of seven satellites because of the electromagnetic pulse generated by space-based nuclear tests in 1962, mutual fear of radiation threats to astronauts, and the risks of nuclear war highlighted by the October Cuban Missile Crisis pushed the two sides to accept military restraint in space out of self-interest. The superpowers didn't reduce their competitive goals in space, but turned that energy in other, less dangerous directions, like human spaceflight and highly secret military support programs for reconnaissance, communications, and early warning of missile launches.

Accordingly, in 1963, they created a kind of sanctuary. Joined by the United Kingdom, Moscow and Washington signed the Partial Test Ban Treaty (PTBT), which banned further nuclear tests in orbit, as well as at sea and in the atmosphere. They also led efforts at the United Nations to pass resolutions in the fall of 1963 calling upon countries not to station weapons of mass destruction in orbit and to accept the extension of international liability law into space. Perhaps most important, they drew on the recently ratified Antarctic Treaty to ban national claims of territory on the Moon and other celestial bodies. The resolutions also urged all countries to provide aid to astronauts in distress in space or those landing outside their home territories. Meanwhile, Western compromises on the UN space committee's membership (adding several Eastern bloc countries) set the stage for the first meetings of that body.

The limited test ban and these basic guiding principles began to create at least a partial governance structure for space activity where none had existed before. To verify these accords, the United States launched the first two space-based nuclear test detection spacecraft (called *Vela Hotel*) in 1963. While it seemed to the public that the United States was still "behind" the Soviets in space, the U.S. military had already begun a tradition of staying one step ahead of the Soviet Union in technical programs critical to national defense. But nobody knew about it for decades.

In the commercial space arena, nations agreed in 1963 to give the International Telecommunications Union (ITU) the right to allocate radio-frequency spectrum for satellite transmissions. Through this action, they headed off potential disputes over satellite communications before conflict had even arisen. The ITU eventually took responsibility as well for the distribution of the limited number of slots over the equator in geostationary orbit, the ideal location for broadcasting and communications satellites focused on a particular region.

Despite these new international accords, tense political rivalry for international leadership in space stimulated both competitive national spending and technological development with the aim of beating the other side to various milestones. From *Sputnik* to Gagarin to Valentina Tereshkova (the first woman in space), the Soviet Union led the early civil space competition. Ironically, arms control and other cooperative mechanisms had reduced the risk of military conflict in a way that made civilian competition safer.

After President Kennedy's assassination in November 1963, his successor, Lyndon Johnson, strongly supported NASA's work and pushed Congress to fund its efforts fully. This commitment allowed continued progress, even in the face of the tremendous technical and occasional political hurdles that threatened to slow or derail the U.S. program. As Catholic priest L. C. McHugh summarized the views of opponents to the Moon shot: "Why should we waste such vast treasures on a remote bit of cosmic real estate, when there is so much to be done on earth?"[12]

But McHugh admitted the minority status of this view in the face of Cold War politics, saying: "Even the taxpayer seems willing to 'go for broke' in the moon race."[13] In 1963, NASA's budget jumped by an amazing 125 percent.[14] By 1966, it rose to 4.4 percent of the federal budget—a level it has not attained since (it is now less than 0.5 percent).[15]

The Soviet Union's weaker financial base and the unexpected death of chief designer Sergei Korolev in January 1966 caused Moscow's Moon effort to weaken considerably. The Soviet program's major technical investment in the unproven N-1 rocket (with a highly complicated system of thirty first-stage engines) also proved to be a fatal mistake.[16]

Despite this competition (table 2.1), trends toward establishing a firmer basis for space's peaceful management proceeded further. By 1966, the United States had submitted a draft space treaty to the UN space committee based closely on the wording of the 1963 UN resolutions. Part of the U.S. desire was to prevent near-term conflict over the Moon, which officials feared might erupt after the first human lunar landing. Despite some initial Soviet objections regarding U.S. overseas tracking stations, Moscow eventually supported the treaty. The United Nations unanimously approved the Outer Space Treaty in late 1966 and opened it for national signatures in early 1967. Key elements of the international agreement included:

- Space exploration should be "carried out for the benefit . . . of all countries . . . and shall be the province of all mankind." (Article I)
- Outer space and the celestial bodies should "not be subject to national appropriation by claim of sovereignty." (Article II)
- Testing of weapons or the conduct of military operations on celestial bodies (including the Moon) "shall be forbidden." (Article IV)
- Countries should conduct all activities in space and on the celestial bodies "so as to avoid their harmful contamination" and should notify other countries before engaging in any activity that might cause "harmful interference" with activities of others. (Article IX)

TABLE 2.1: U..S. and Soviet Space "Firsts" During the Cold War

Activity	Country	Date	Spacecraft
Earth satellite	USSR	October 1957	*Sputnik 1*
Animal in orbit	USSR	November 1957	*Sputnik 2* (Laika)
Spacecraft landing on the Moon	USSR	September 1959	*Luna 2*
Meteorological satellite	USA	April 1960	*Tiros 1*
Navigational satellite	USA	April 1960	*Transit 1B*
Electronic intelligence satellite	USA	June 1960	*Grab*
Photo-reconnaissance satellite	USA	August 1960	*Corona/Discoverer 14*
Communications satellite	USA	October 1960	*Courier 1B*
Man in orbit	USSR	April 1961	*Vostok 1* (Yuri Gagarin)
Woman in orbit	USSR	June 1963	*Vostok 6* (Valentina Tereshkova)
Nuclear detection sensor	USA	October 1963	*Vela Hotel*
Spacewalk	USSR	March 1965	*Voskhod 2* (Alexei Leonov)
Manned lunar landing	USA	July 1969	*Apollo 11* (Neil Armstrong, Buzz Aldrin)
Venus landing	USSR	December 1970	*Venera 7*
Space station	USSR	April 1971	*Salyut 1*
Mars landing	USA	July 1976	*Viking 1*
Reusable space shuttle	USA	April 1981	*Columbia*
Yearlong human spaceflight	USSR	December 1988	*Mir* (Vladimir Titov, Musa Manurov)

SOURCE: James Clay Moltz, "Managing International Rivalry on High Technology Frontiers: U.S.-Soviet Competition and Cooperation in Space" (Ph.D. diss., University of California, Berkeley, 1989), table 3, "U.S. and Soviet Space Firsts," 67.

- All stations on the Moon or other celestial bodies "shall be open to representatives of other States Parties . . . on the basis of reciprocity." (Article XII)[17]

The treaty soon established itself as the single most important legal agreement affecting space activity. In 1967, Austria, Iran, and Egypt proposed a further step with a proposal to the UN space committee that a global space organization be formed on the model of the International Atomic Energy Agency to both monitor space activity and promote the spread of space technology to less-developed nations.[18] But the proposal met with little support in the midst of the Cold War and the ongoing U.S.-Soviet Moon race. Meanwhile, the increasing risks of the intensifying pace of the lunar race hit hard on both countries' space programs in 1967.

In a January ground test in Florida of the new *Apollo 1* capsule, three U.S. astronauts suffocated as they tried to escape a fire that erupted. Strapped into their seats, they could not free themselves and operate a manual hatch fast enough before the raging fire consumed all of the oxygen in the module. Over the objections of the capsule's manufacturer, NASA had ordered an oxygen-only environment to be used, rather than a less volatile mixed-gas system, as a cost-saving measure.[19] A few months later, the Soviets' first *Soyuz 1* capsule, which survived a problematic orbital mission marred by equipment failures and had reentered the atmosphere en route to landing in Soviet Central Asia, failed to deploy its parachute properly. The capsule's high-speed impact with the ground killed veteran cosmonaut Vladimir Komarov, leading to soul-searching and a stand-down order within the Soviet manned program.[20] Space historian Asif Siddiqi concludes: "The responsibility and guilt for the accident lay . . . on a technological culture that considered high risks acceptable in the cause of satisfying political imperatives."[21] These terrible accidents contributed to the signing of the Agreement on the Rescue of Astronauts, the Return of Astronauts and the Return of Objects Launched into Outer Space in 1968, which required the two sides to repatriate all spacecraft that landed outside their home country and to provide assistance to one another's spacefarers in case of accidents in space or upon their return to Earth. But the lack of a compatible docking mechanism among the programs' spacecraft meant that implementation would have to wait for a thaw in political *and* technical relations.

The United States continued to struggle with the von Braun–led Saturn V launcher needed to send its astronauts to the Moon. In April 1968, a test launch with an unmanned *Apollo 6* capsule aboard suffered from a "pogo effect." Unnerved NASA engineers watched as their thirty-six-stories-tall rocket bounced across the pad for half a minute before finally achieving liftoff.[22] It had to be with considerable intestinal fortitude that astronauts Walter Schirra, Donn Eisele, and Walter Cunningham entered their *Apollo 7* spacecraft in October for the system's first crewed flight. Fortunately, the mission succeeded, setting the program on course for the three follow-on flights needed to set up the remarkable lunar landing by *Apollo 11* on July 20, 1969. A desperate Soviet effort to land an unmanned rover on the Moon on the eve of the *Apollo 11* mission ended in failure as the spacecraft exploded seconds after takeoff. After this devastating loss, the Soviet Union abandoned its efforts to land cosmonauts on the Moon and regrouped, focusing more attention on space stations, long-duration human spaceflight, and military programs.[23]

The tremendous world attention paid to the space race and the lure of orbital flight had by now begun to attract other countries. Like the United States and the Soviet Union, France had gained a head start on advanced rocketry after World War II from German scientists who had worked on the V-2 program.[24] In the 1950s and early 1960s, France developed a suborbital sounding rocket program with the aim of eventually launching a satellite. After testing its first nuclear weapon in 1960, France sought to develop ballistic missiles to deliver nuclear bombs, which would also give it space-launch potential. In 1965, France's three-stage liquid- and solid-fuel Diamant rocket succeeded in delivering the *Asterix* satellite into orbit from a launch site in Algeria, making France the world's third country to enter the space age.

In Asia, Japan's economic recovery in the 1950s and 1960s stimulated an interest in developing space capabilities, for both scientific and economic purposes.[25] In order to prevent any possible drift toward military applications, the national legislature, the Diet, passed a resolution in 1969 explicitly barring the country from engaging in military space activities. In January 1970 Japan succeeded in becoming the fourth country to launch its own satellite (*Ohsumi*) aboard its domestically built solid-fuel Lambda 4S-5 rocket.[26] Japan's scientists and engineers soon began to cooperate more closely with the United States and acquired assistance in liquid-fuel rocket technology in order to build a more powerful next-generation launch system.

China took a more typical military-led route into space. Following the Communist revolution in 1949, the young Chinese government received considerable technical assistance from the Soviet Union, including prototypes of two early Soviet rockets (the V-2-derived R-1 and R-2) in the late 1950s.[27] But the political split between the two capitals in 1960 ended this cooperation, creating a crisis for China's fledgling space efforts. Many Chinese space scientists and engineers would soon fall victim to the Cultural Revolution, further setting back the program. But support from China's military leaders, who were eager to build a delivery system for China's atomic bomb (first tested in 1964), saved the rocket effort and opened the possibility of spaceflight.[28] Finally, in April 1970, China became the fifth country to launch a satellite into orbit. Yet China's leadership struggles and on-again-off-again support for space research meant that the country had a long way to go before it could carry out a normal space program.

The United Kingdom joined the club in October 1971, launching the *Prospero* satellite from a Black Arrow rocket in Australia. But it soon canceled its launch program because of cost concerns and access to U.S. launchers.

In the U.S.-Soviet relationship, meanwhile, new tensions over Soviet deployment of nuclear-tipped anti-missile defenses around Moscow and U.S. plans to match them called emerging nuclear parity into question. Late in the Johnson administration, the United States started talks with Moscow aimed at limiting missile defenses in order to prevent a defensive technology race or, alternatively, a requirement to build more offensive missiles to overcome the new defenses. But the Soviet Union proved initially unreceptive.

DÉTENTE IN SPACE

A détente relationship gradually emerged between the new Richard M. Nixon administration and the Brezhnev-Kosygin leadership in Moscow out of shared interests. In the post-Apollo environment, neither side could afford the massive expenditures that had characterized the Moon race, and both wanted to avoid the possibility of a new competition in anti-ballistic missile defenses, given the benefits of stable nuclear deterrence. For space, these developments resulted in: (1) a scaling back of civil space competition and instead beginning plans for a *Soyuz-Apollo* spacecraft docking; and (2) unprecedented cooperation in space-related arms control. The rapprochement in the military sphere included a ban on testing or deploying space-based missile defenses in the 1972 Anti-Ballistic Missile (ABM) Treaty and also formal agreement not to interfere with each other's "national technical means" of verification (i.e., reconnaissance satellites). These new rules created greater safety and predictability, while not ending competition altogether.

At the multilateral level, the United Nations passed the Convention on International Liability for Damage Caused by Space Objects in 1972. This document clarified the procedures called for in the 1963 UN space resolution on legal principles and in the 1967 Outer Space Treaty. It stipulated that claims must be submitted within a year following the date of any incident and that the process of resolution would first involve bilateral discussions and, if those were unsuccessful, the formation of an international claims commission headed by a neutral third party.[29] This treaty provided the legal basis for claims by Canada in 1978 when a Soviet radar satellite powered by an onboard nuclear reactor crashed in Canadian territory and spewed hazardous highly enriched uranium across a wide area. Although the Soviet Union never admitted its guilt, Moscow eventually paid a settlement of $3 million toward the cleanup costs.[30]

International efforts to begin keeping better track of objects in orbit emerged in the 1975 UN Convention on Registration of Objects Launched into Outer Space. This agreement required launching countries to provide information for a long-planned UN registry with the name (or number) of the spacecraft, the date and location of its launch, its initial orbital parameters, and its general function in space.[31] The registry became the single accessible source of information on spacecraft, although adherence to the terms remained imperfect, with countries often covering military purposes with a "scientific" mission statement or providing incomplete or late information. Nevertheless, the Registration Convention established itself as a key tool for countries in identifying and tracking space objects. In recent years, it has been supplemented by online spacecraft tracking lists populated with information provided by amateur astronomers and scientists.

In the face of the vast scale of U.S. and Soviet space activities, European countries recognized that they could not hope to compete or develop viable space industries if they continued to pursue purely national programs. After experiencing problems in joint efforts at satellite development under the European Space Research Organisation and in joint booster work under the European Launch Development Organisation in the 1960s, Western Europe's leading space actors began the process of creating a more integrated and comprehensive space organization in 1971.[32] The European Space Agency (ESA), formed in 1973, established a system of both divided and shared responsibilities and funding. France took the lead in building the planned Ariane launcher and agreed to pay for the lion's share, while West Germany announced that it would bear the primary responsibility for scientific modules that would eventually fly on the U.S. space shuttle. Britain agreed to provide maritime communications satellites.[33] The ESA's first independent satellite launch finally took place in December 1979 with the successful release of a scientific satellite by France's Ariane 1 rocket.

Despite advances in international space cooperation and the conduct of the landmark 1975 *Apollo-Soyuz* spacecraft docking, however, the U.S.-Soviet détente relationship began to cool. Washington and Moscow clashed over policies in the Third World and with regard to nuclear arms, as well as renewed differences over space weapons. The Soviet Union resumed testing of a new version of its kinetic anti-satellite (ASAT) system—a radar-equipped homing device that maneuvered into the same orbit as its target and then exploded nearby[34]—causing the United States to pull back on civil space cooperation. One of the last acts of President Gerald Ford's administration was to initiate a U.S. ASAT research and development program to counter the Soviet military's effort.

Meanwhile, the series of Moon landings by NASA astronauts and by unmanned Soviet probes and rovers had caused concern among other nations that the two superpowers might establish permanent settlements on the Moon's surface and begin to derive untold wealth from mining or other activities there. As a result, the United Nations began to negotiate a treaty that would attempt to prevent lunar conflict and develop a potential mechanism for managing access to and benefits from lunar resources.[35] These talks drew on the agreed language in the Outer Space Treaty stating that space is "the province of all mankind" and that exploration "shall be carried out for the benefit and in the interests of all countries, irrespective of their degree of economic or scientific development." The eventual Moon Treaty, finalized in 1979, proposed the creation of an international organization that would be responsible for allocating resources and profits to ensure that all countries benefited from lunar exploration. The treaty language called upon states to accept an "equitable sharing" of benefits that would take into account the "interests and needs of the developing countries."[36] The agreement did not spell out exact percentages of national profits versus shared benefits. Perhaps not surprisingly, the main spacefaring countries declined to support the new treaty. This inaction raised questions about how future lunar activities would be governed.

Among the new actors who shared a skeptical view of the superpowers' intentions in space was India. Although India had received NASA training, ground stations for Landsat remote-sensing data, and technological assistance in the form of Nike-Apache rockets in the 1960s, it remained wary of dependence on Washington. It therefore sought to balance U.S. assistance by seeking Soviet technical aid. As part of a deal granting the Soviet Union access to Indian ports to service its space-tracking ships in the early 1970s, Moscow launched India's first domestically produced satellite in 1975 and provided additional technical training to Indian scientists.[37] India eventually developed its own space-launch vehicle and, in 1980, succeeded in delivering the small *Rohini 1* communications satellite into orbit. Within a few years, the Soviet Union trained India's first cosmonaut for a flight aboard the *Salyut 7* space station in 1984.

Back on the Precipice of Space Conflict

In the late 1970s, U.S. fears surrounding the resumption of Soviet ASAT testing led the Carter administration to engage Moscow in negotiations concerning a possible treaty against ASAT weapons.[38] The participating officials struggled to make progress, however, given political and strategic differences, including Moscow's reluctance to give up the advantage it saw in its ground-launched, co-orbital ASAT interceptor. Definitional questions also plagued the talks, with Moscow insisting that the planned U.S. space shuttle be included in any ASAT ban because of its expected capability to pluck satellites from orbit for repair. In the end, the two sides came close to a ban on the use and testing of ASAT weapons, which was more easily verifiable than a prohibition on ASAT possession. But continued mutual distrust, the U.S. loss of its eavesdropping base in Iran (as a result of the 1979 Iranian revolution), and the hostile politics surrounding the Soviet invasion of Afghanistan that December caused the proposed compromise to be pushed to the back burner.

The 1980s yielded no new space security agreements. High tensions surrounding the growing Soviet nuclear arsenal and the U.S. decision to deploy new weapons in Europe as a countermeasure poisoned the atmosphere for talks. Nevertheless, new Soviet proposals emerged during this period, including a perhaps self-serving 1981 draft treaty to prohibit new ASAT tests and ban threats against space objects, hoping to block planned U.S. ASAT testing. In the face of clear U.S. commitment to move forward with ASAT and new anti-ballistic missile research, the Soviets announced a unilateral test moratorium and proposed a new draft treaty banning all "space-strike" weapons and the destruction of existing ASAT systems.[39] But U.S. fear of possible Soviet cheating quashed the effort, and the U.S. Air Force went forward with an eventual kinetic test of an aircraft-launched ASAT missile in 1985. Still, the two sides continued to talk about space security in the context of the Defense and Space Talks, part of the ongoing U.S.-Soviet arms control process.

U.S. military analysts worried increasingly about Soviet deployments of large, multiple-warhead SS-18 nuclear-armed missiles and the risk of a possibly disarming Soviet first strike. While critics discounted this scenario, supporters managed to convince President Ronald Reagan in 1983 to adopt a program called the Strategic Defense Initiative (SDI) aimed at stopping the SS-18s in their initial boost phase. SDI, derided by critics as "Star Wars," would have required the deployment of thousands of space-based interceptors in order to accomplish its mission, as well as U.S. withdrawal from the ABM Treaty.[40] In the end, further research showed that ambitious scientists had oversold the technology, and the costs were far too high. But politics ended up making a greater difference.

The emergence of General Secretary Mikhail Gorbachev changed the Soviet Union and its space policy fundamentally. Gorbachev canceled various Soviet space weapons programs, reduced competitive civil space efforts, and called for the creation of a world space organization to replace Cold War competition. While the United States was skeptical of the scope of Soviet changes, it agreed to renew U.S.-Soviet space cooperation in 1987 with deep-space missions and a set of working groups to explore additional initiatives.[41] The SDI program, meanwhile, ended up being limited to research and development, although some technologies were folded into later U.S. missile defense programs.

Israel became the next country to join the space club with its successful launch of the *Ofeq 1* reconnaissance satellite aboard a Shavit rocket in 1988. Israel had obvious national security needs for space-based reconnaissance and, like those of other spacefaring countries with nuclear weapons, its space efforts benefited from its missile delivery system.[42] But Israel was hampered by an unfavorable location, which forced it to launch westward against Earth's orbital direction (requiring more fuel) or eastward over rival countries that might intercept its technology in a failed launch. Israel would eventually develop unique expertise in miniaturization and a partnership with India to launch its increasingly sophisticated military reconnaissance satellites.

EARLY POST–COLD WAR COOPERATION

After the Soviet Union's breakup in late 1991, budget pressures in both Russia and the United States led to the forging of much deeper cooperative ties. As the United States sought to recover from Reagan-era deficit spending, it found major space initiatives like the proposed space station unaffordable, despite cooperation with several U.S. allies. Washington also worried that economic difficulties in the Soviet successor states would cause impoverished space scientists and engineers (especially in Russia and Ukraine) to sell their know-how to would-be missile proliferators in Asia and the Middle East.[43] Similar fears drove the United States to develop the so-called Cooperative Threat Reduction Program in the nuclear sector. In the end, these mutual interests created an unprecedented agreement between President Bill Clinton and President Boris Yeltsin to include Russia in the revised *International Space Station* (*ISS*). NASA provided some initial funding for astronaut visits to the Soviet *Mir* space station, while purchasing Russian technology to develop new options for its military and commercial space launch needs. The *ISS* would also benefit from Russian-provided modules and supply boosters. Ukrainian space scientists had fewer options, but some benefited from a joint venture called Sea Launch formed by a U.S.-Ukrainian-Russia-Norwegian consortium organized to provide mobile launch services from a refitted barge at sea, allowing favorable equatorial sites to be used.

As Russia's space sector reoriented itself on a commercial basis, its military space program experienced a near collapse from a lack of government contracts. In the 1990s, Moscow suffered through a decade of hyper-inflation, low tax returns, capital flight, lawlessness, and inadequate investment.[44] Despite some delays, however, the Russians launched the first module for the *ISS* from Kazakhstan in November 1998. This spacecraft soon docked with the first U.S.-launched element and construction spread over the next thirteen years.

Although U.S.-Russian civil space cooperation expanded in the 1990s, space arms control activity fell off the radar screen. In this new international environment, the United States saw no real competitors. The perceived low demand for new space security agreements under the Clinton administration, combined with stiff opposition by Senate Republicans to any new space-related treaties (for fear of limiting U.S. missile defense options), effectively silenced U.S. space diplomacy. Instead, the United States used consensus rules at the Conference on Disarmament in Geneva to block international efforts to restart negotiations on space arms control and the prevention of an arms race in space.[45] After the George W. Bush administration entered office in 2001, it removed one of the former keystones of space security by announcing that it would withdraw from the 1972 ABM Treaty in order to open the path to nationwide missile defenses. In space, this move cleared the way for the possible deployment of space-based weapons. Objections by new Russian President Vladimir Putin and even by a coalition of U.S. allies failed to halt this shift. The Bush administration funded research into space-based interceptors, but none would be tested or deployed during the Bush years because of their high cost and the immaturity of the technology.

During the period from 1979 to 1998, the United States and China cooperated in civil and commercial space activities. The initial motivation for this collaboration had stemmed from a common mistrust of the Soviet Union. Two Chinese experiments had flown on a shuttle mission in the early 1990s, and the United States had opened the market for U.S. commercial satellite launches on Chinese rockets. But changes in congressional instructions on export controls in 1999 brought this to a halt by recategorizing all U.S. space technologies as munitions items. These changes came about because of the conclusions of the congressional Cox Committee tasked with investigating charges of Chinese spying, mainly against U.S. nuclear facilities but also in the missile sector.[46] A new presumption to deny space exports to foreign entities caused U.S. satellite sales to plummet, as even U.S. allies found the procedures for waivers onerous.

More-stringent U.S. export controls initially had the desired effect of stemming the possible outflow of military secrets to China and "punishing" its commercial launch sector. Within a few years, however, other countries without these limitations stepped into the breach to begin engaging in commercial space cooperation with China. When tightening export controls, the U.S. Congress had neglected to consider the wide availability of satellites and other space-related technologies on the international market. Russia, Ukraine, Germany, France, Italy, and the United Kingdom all benefited from the new U.S. regulations, and many U.S. allies simply stopped including U.S. technologies in their spacecraft in order to avoid U.S. restrictions and promote their own export industries. As a result, both civil and commercial ties between these countries and China strengthened, while the United States became increasingly isolated from cooperative activities with Beijing in space. These developments highlighted how much the space economy had changed since the Cold War, when the United States could act with virtual impunity in controlling the commercial space market.

CHALLENGES OF TWENTY-FIRST-CENTURY SPACE POLITICS

The current period in international space politics began with China's successful launch of its first *taikonaut* (astronaut) on *Shenzhou 5* in October 2003, bringing to fruition plans that had been made in the 1980s.[47] In addition, China expanded its capabilities across the board in other space missions: scientific, commercial, and military. China's strong political and financial commitment to space activity marked the beginning of a renewed contest for space supremacy. Although the United States still maintained a significant lead, China's 2007 test of an anti-satellite weapon placed the space assets of both the United States and other countries at risk. The U.S. decision to modify a sea-based anti-ballistic missile system to destroy an ailing intelligence satellite in 2008 showed that it, too, maintained the capability to attack space assets if need be, despite mothballing its earlier F-15-based system.

The actions of 2007 and 2008 forced other spacefaring nations to recognize the importance of space activity to modern defense capabilities, particularly reconnaissance, navigation, and communications. While much information on one's neighbors or rivals could be purchased from international Earth-imaging companies (such as the French SPOT or the U.S. DigitalGlobe), some national militaries wanted greater control over the imagery and assurance that such access would not be cut off in a time of conflict. Space technology was also becoming increasingly critical to regional and global military force management, including for countries engaged in either UN support operations or power projection abroad. Moreover, space offered traditional benefits like prestige, which could contribute to national power in intangible ways.

Iran's decision to develop a space program included all of the above elements. After it established a space agency in 2004, Iran conducted a series of suborbital flights with sounding rockets, some with live animal payloads. In February 2009, Iran succeeded after several attempts in orbiting its *Omid* satellite aboard a Safir rocket, making it the second country in the Middle East to accomplish this relatively complex technical feat. Perhaps optimistically, Iran announced that it would develop its own human spaceflight program. Not surprisingly, Turkey and other regional rivals took notice and developed plans to expand their space activities as well.

Similar regional dynamics were clearly under way in Asia. China's rapid surge in civil, commercial, and military space activities challenged its main regional rivals for influence and power: Japan and India. After 2003, Japan responded by reorganizing its civil space program into the Japanese Aerospace Exploration Agency (JAXA) and conducting an impressive array of sophisticated scientific missions.[48] Tokyo also recommitted itself to human spaceflight with a series of missions to exploit its *Kibo* module on the *ISS*. Finally, in 2008, the Japanese Diet rewrote the national legislation governing space activities to allow military programs for the first time.

India similarly revamped its space program in reaction to China's sudden rise and the political, economic, and military threats it perceived in orbit. It abandoned the country's long-held, singular focus on space applications to serve its large population and initiated several high-prestige space science missions and a major move into military activities.[49] Indian officials also declared their intention to develop ASAT technologies to match those of China.

Spurred by deep-seated political rivalry and emerging Asian space developments, North and South Korea began to engage in space competition of their own. Benefiting from its active missile program, North Korea first attempted a satellite launch in August 1998, but the satellite stage failed and the spacecraft crashed into the Pacific Ocean. South Korea responded by accelerating its space efforts, including the purchase of foreign satellites to expand its capabilities, the development of a satellite manufacturing infrastructure, and the establishment of an Earth-remote sensing center for monitoring East Asia and especially North Korean military sites. South Korea also constructed a launch facility and pledged to develop an independent space booster. North Korea inaugurated a new launch site of its own in 2012, and orbited its first satellite in December of that year. After several failures, South Korea responded by launching a more sophisticated satellite into orbit in January 2013.

Increasing competition meant more countries and more objects in space. All of these activities over-stressed Cold War–era space management mechanisms, particularly as crowding was now exposing the surprising "finiteness" of near-Earth space. As more nations wanted satellite communications, there were too few slots available now in geostationary orbit. The radio frequency spectrum was nearly fully allocated in the most desirable bands, creating logjams for countries and companies seeking to offer new services. Militaries were pushing the boundaries of "peaceful uses" and avoiding the Outer Space Treaty's requirement for consultations in advance of harmful activities by claiming that their activities were defensive and not subject to the somewhat vague requirements treaty.

But cooperation and commercialization were also proceeding apace. Space business was at record levels and promised to rise even higher in the next decade, thanks to the emergence of more direct-to-home and direct-to-device services, as well as new and cheaper forms of space tourism. Many scientific missions were now conducted through international cooperation, both to reduce costs and to increase the prospects of discovering something new through the sharing of additional sensor technology. In May 2008, the *ISS* had reached its full complement of six astronauts for the first time, and all major construction was successfully completed in 2011 after the final U.S. shuttle flight. High costs made it likely that future lunar missions might be collaborative as well. Still, political tensions—particularly between the United States and China—made such comprehensive cooperation seem difficult to reach.

CONCLUSION

In the current history of space politics, all countries active in space say that they are complying with the 1967 Outer Space Treaty and are engaged in only "peaceful purposes." But what that means and where international space relations are headed are becoming far more complicated issues. An increasing number of national militaries see space activities as critical to their future capabilities in peacetime and in war, creating tensions between neighbors and raising the stakes for existing rivalries. That said, the expansion of new commercial actors in space, which are often transnational in their ownership and funding, complicates old patterns by making profit rather than power an even more important objective. Similarly, the lowering of the threshold of space activity with the advent of micro-satellites, secondary payloads (free launches for small satellites as spare cargo), and cheaper commercial boosters suggests that academic, nonprofit, and scientific actors will have an increasing stake in space. Finally, the likely rapid growth of commercial human spaceflight in the next decade will put a new premium on preserving safe access to space, possibly constraining national militaries. How these conflicting agendas will interact in the face of technological advances and new efforts at both space rivalry and cooperation will be explored in the next several chapters.

3

CIVIL SPACE
Science and Exploration

Space exploration offers the potential for human beings to learn more about *why* we are here on Earth, *where* the building blocks of life might have come from, and *what* our future as a species might be. Understanding outer space better is also critical for beginning to open the option for people to live away from Earth on a permanent basis. Such a capability may be essential for human survival in case of the ruination of our planet by a man-made or natural disaster. In the popular 2008 movie *WALL-E*, for example, humans abandon Earth because of its contamination from centuries of accumulated garbage and pollutants, leaving only robots to clean up the mess. Under-exercised, blimp-like humans are stranded in space, but they eventually begin to return to Earth. While not exactly an ideal scenario, a human ability to live for an extended period in space proves essential to our species' survival.

Yet the *WALL-E* story requires international cooperation to construct Earth's orbital space station and to unify people behind the mission. By contrast, in the past, forces of international *competition* have been the major motivation for countries to support space exploration efforts, such as the Moon race in the 1960s. Historically, exploration of new continents on Earth led to commerce, which led to conflict, which led to warfare. Thus far, however, space exploration has not led to such conflict, perhaps because of the limits of space commerce, the ban on territorial claims, and the lack of permanent settlers. Since the end of the Cold War, there have been some emerging differences in the way space science and exploration have been conducted. Many major spacefaring countries are now cooperating in human and robotic exploration, with one major exception: U.S.-Chinese space relations. A central question about future civil space activities—such as human missions to Mars or major robotic missions to distant planets—is this: are new forms of cooperation likely to emerge, or are past competitive practices likely to reassert themselves? In the late 1950s, the British space writer Arthur C. Clarke captured this duality well. He saw the potential for great tragedy in space because of the nuclear competition of the Cold War, observing, "I am not unmindful of the fact that fifty years from now, instead of preparing for the conquest of the planets, our grandchildren may be dispossessed savages clinging to the fertile oases in a radioactive wilderness."[1] But he went on to capture the potential for a great cooperative breakthrough as a result of space activity, arguing, "The crossing of space . . . may do much to turn men's minds outwards and away from the present tribal squabbles. In this sense the rocket, far from being one of the destroyers of civilization, may provide the safety valve that is needed to preserve it."[2]

"Civil" spaceflight means nonmilitary and noncommercial activities in orbit and in deep space, typically paid for by governments. This category has traditionally included space science, robotic exploration, and human spaceflight. To date, national governments have played the leading role because of the high costs involved, the limited means of other civil organizations, such as universities, private groups, or international bodies, and the lack of obvious commercial benefits. Instead, the main value of space science and exploration has come in the form of knowledge and, occasionally, political prestige.

Space science encompasses such high-profile activities as planetary exploration (like NASA's Voyager missions and the Mars Rovers), heliophysics (research about the sun and space weather), astrophysics (research about matter, gravity, and the universe), and Earth observation and monitoring (data collection on such matters as ocean salinity, climate change, and weather prediction). Besides NASA and the Russian Space Agency, the European Space Agency (ESA), Japan, and several other countries have conducted unique missions that have yielded important new findings about and from space.

Countries that have followed the leaders in space science may not be able to claim scientific or technological "firsts," but they may still succeed in gaining domestic and international benefits. China's human spaceflight program represents such an instance. Despite the lack of any new scientific value in its manned missions thus far, the Chinese government has reaped major benefits in domestic prestige and foreign respect. The same could be said—to a lesser extent—of recent missions by Japan, China, and India to map and image the Moon.

As mentioned, space science, robotic missions, and human exploration have traditionally been conducted through national space programs, which have borne the expenses, taken the risks, and claimed the credit. In many cases, these projects have exacerbated rather than reduced international tensions, as one country gained new influence and respect relative to its rivals. But major scientific missions are very expensive, which makes their selection, development, and implementation a fundamentally political activity, given that the public is paying the bill. In periods of great political rivalry, these resources are often easier to come by because of the centrality of such expenditures for, in the words of the historian Walter McDougall, "the survival of national self-image."[3] These pressures, he argues, created the drive toward "technocracy" in the late twentieth century, or the institutionalization of state-led sponsorship of science and technological development.[4] Governments that are authoritarian (such as China, Iran, and North Korea today) have an advantage in that they do not have to ask their populations to approve costly expenditures on space. But political change can jeopardize these conditions, as the Russian space program learned to its chagrin after the end of Communist rule in the 1990s.

The high cost of major space missions means that this paradigm has begun to be challenged by the increasing frequency of cooperative international missions. Examples include the *International Space Station*, the NASA-ESA *Cassini-Huygens* project, and the French-Indian *Megha Tropiques* Earth observation satellite. As Jim Green, the director of NASA's Planetary Science Division, argues about the changing conditions affecting space science, "We need to partner now more than ever before to be able to make those [expensive] missions a reality."[5] At the same time, international cooperation may pose a risk if a foreign partner backs out. Such projects sometimes fail entirely, leading to heightened international tension. Cost-sharing sometimes fails as well, leaving one partner holding the bag. As the space analysts Roger Handberg and Zhen Li observe, "Large-scale space projects can turn into financial albatrosses."[6] Thus, while collaboration is sometimes smart, successful collaboration is not easy.

In order to capture current trends in civil space, it is useful to begin with a brief survey of the major national programs and their main directions, projects, and funding, as well as their foreign partners. Clearly, it will make a difference if countries desire to conduct purely *national* space programs or if they see a need to engage in joint planning, cost-sharing, and in-depth operational collaboration. One reason may be growing appreciation for the shared environmental risks that all countries face in space, including dangerous near-Earth objects (asteroids), possible contamination of Earth by organisms from space, and harmful orbital debris. In space, it does not take much reflection to realize Earth's precarious place. The scientist Donald Yeomans describes risks of catastrophic asteroid damage by soberly noting that "Earth runs its course around the sun in a cosmic shooting gallery—with us as the target."[7] A relatively small meteor, 50 feet in diameter, exploded with a force equivalent to more than thirty-eight Hiroshima-sized nuclear bombs about 12 miles above Chelyabinsk, Russia, in February

2013, injuring more than a thousand people and damaging hundreds of nearby buildings.[8]

Recognizing this mutual vulnerability, scientists typically are the most cooperative of all participants in space activity. A recent report from a group of international scientists observed, "Science has the power to act as a bridge between spacefaring nations and other stakeholders and the ability to engage society and promote participation while delivering direct benefits to the public."[9] But scientists are often funded by governments, whose officials may have other ideas about the desirability of cooperation with countries they consider to be rivals, setting up a dilemma. Examining the trade-offs involved in these choices is critical to understanding the likely direction of civil space activities and whether they are likely to lead toward future cooperation or instead competition.

LEADING CIVIL SPACE PROGRAMS AND PLANS

Historically, only a few countries have possessed the capability to carry out independent space science missions, and even fewer have been involved in human spaceflight programs (table 3.1). The civil space programs of these countries are likely to shape the coming few decades in space and have a major influence on the character of international relations in orbit. The most important organizations include NASA, the European Space Agency, the Russian Space Agency, the Japan Aerospace Exploration Agency, the similar-sounding China Aerospace Science and Technology Corporation and the China Aerospace Science and Industry Corporation, and the Indian Space Research Organization. It is worthwhile to review each, including their histories, domestic politics, and main areas of cooperation with other space programs.

NASA

As the world's leading space agency, NASA is important not only for what it does in space but for how its programs influence others. It is involved in a large variety of cooperative initiatives. But the Obama

TABLE 3.1: World's Leading Space Programs

Country	Civil space budget (in yearly U.S. $)	Total space industry personnel	Total launches per year (average)
United States	$16.5 billion	250,000	15
European Space	$5.6 billion	34,000	8
Russia	$4.4 billion	250,000	29
Japan	$3.8 billion	7,000	3
China	$3 billion*	300,000*	19
India	$1.5 billion	32,000*	3

*Estimated figure.

Sources: Peter B. de Selding, "Additional Funds, New Members Boost ESA's 2013 Budget," *Space News*, January 28, 2013, 5; "Budzhet na 2013 god predpolagaet rekordnoe finansirovaniye kosmonavtiki" (Record financing proposed for 2013 space budget) (Roscosmos press release, in Russian, June 6, 2012); Dan Leone, "After Budgetary Dust Settles, NASA Left with $16.5B for 2013," *Space News*, March 15, 2013, 4; Space Foundation, *The Space Report: 2012* (Colorado Springs: Space Foundation, 2012), ch. 2, 3, and 4; personnel and launch figures include civil and military activities; figures for launches per year are averaged from 2011 and 2012.

administration's decision to abandon the costly Constellation program that envisioned an international human return to the Moon and NASA's declining budget have created uncertainties concerning future U.S. leadership. Still, NASA has a long legacy of successful space science missions and remains the most desired partner for many nations.

Historically, NASA's flagship projects have involved high-cost, U.S.-designed and -built spacecraft, based on cutting-edge science, literally going where no one has gone before. In the 1960s the Moon landing marked the main highlight, but robotic missions such as the *Viking* landers on Mars in 1976, the *Voyager* spacecraft that explored Jupiter, Uranus, Saturn, and Neptune in the late 1980s, and the *Magellan* lander that reached Venus in 1991 featured prominently in building a broader base of knowledge about our solar system. The *Hubble Space Telescope*, launched in 1990, vastly expanded our understanding of worlds beyond the reach of previous land-based telescopes. Its images sparked public interest in astronomy and in existential questions about Earth's place in the universe, which came to be seen as but one planet in one solar system amidst a myriad of now-visible others.

With the exception of the last of these projects (*Hubble*), NASA's major exploratory efforts during the Cold War typically represented U.S.-only efforts with very limited participation by foreign space programs. For political reasons, we had relatively little cooperation with our space peer (the Soviet Union) during the Cold War, and the capabilities of our allies remained far behind us in most fields. But there were some important exceptions, such as the remarkable *Apollo-Soyuz* docking in 1975,[10] and a series of data exchanges and cooperative missions during this time, particularly in space biology. These contacts built significant linkages between the two sides' scientific communities. While hostile high-level politics nixed these programs after 1979, they eventually returned and expanded following Soviet political reforms in the late 1980s. Cooperation with Russia has grown steadily since that time, both because of cost concerns and in order to benefit from its technologies and experience, especially in human spaceflight.

Space science cooperation with U.S. allies can be traced deep into NASA's history, The National Aeronautics and Space Act of 1958 stated that NASA "may engage in a program of international cooperation" to serve U.S. foreign policy interests.[11] NASA quickly began reaching out via the International Council of Scientific Unions, which formed a Committee on Space Research (COSPAR).[12] But COSPAR lacked the mechanisms for functional cooperation with a government agency, and NASA turned to government-to-government agreements, beginning discussions with the United Kingdom in 1959, resulting in the U.S. launch of Britain's *Ariel 1* satellite in 1962. Later, NASA worked with a broad swath of European countries by reaching an agreement with the European Space Research Organisation. It also cooperated with India and Brazil by the mid-1960s, and later with Japan.

Arguments for international space cooperation gained greater strength in the 1970s as NASA's budget declined and the political purposes of space activity took on a new dimension: involving allies in more advanced technical endeavors. In the early 1970s, NASA engaged

U.S. allies in several projects linked to the planned U.S. space shuttle, although not without disagreements over priorities.[13] During the process of constructing the planned *Freedom* space station in the late 1970s and early 1980s, NASA recognized that it would have to attract both supporting finances and technology (including modules) from its main allies with spacefaring capabilities: the newly reconstituted European Space Agency, Japan, and Canada. But even with this support, the space station project faced potentially debilitating cost overruns by the late 1980s, well before any hardware had even been launched. Benefiting from a cooperative human spaceflight agreement with Russia signed at the end of the George H. W. Bush administration in October 1992, the Bill Clinton administration crafted legislation in 1993 to provide $400 million in contracts to the newly formed Russian Space Agency (later known as Roscosmos) to benefit from its technology, save money, and gain a valuable new partner.[14] This out-of-the-box approach nearly caused the U.S. Congress to deny the funding for the project altogether (it passed in the House of Representatives by one vote). The new deal saved the renamed *International Space Station* (*ISS*).

But the station's construction could be more accurately described as "collaboration" rather than truly *joint* development because of the complexities and sensitivities involved in sharing national space technologies. As one analyst described U.S. goals in space station cooperation in the late 1990s, "Ideally, the engineering interface between US and foreign elements would be kept to a minimum."[15] That is, while getting the national modules of the *ISS* to fit together required a great deal of coordination, as did the delivery of supplies and astronauts, very little of the station's hardware was built in a cooperative fashion by member states.

With the building blocks of the *ISS* in place and its operation under way, the United States proposed a follow-on cooperative plan for space in January 2004 called the Vision for Space Exploration (VSE), announced by President George W. Bush in a speech at NASA headquarters. The concept outlined a human return to the Moon by 2020 and the use of a lunar station to prepare for the first manned mission to Mars. The administration proposed that part of the several hundred billions of dollars the VSE would cost would come from partner nations. As President Bush noted, "The vision I outline today is a journey, not a race, and I call on other nations to join us in this journey, in a spirit of cooperation and friendship."[16] Facing the rising costs of the wars in Afghanistan and Iraq, the American public proved largely unsupportive of such an expensive space endeavor (as indicated by opinion polls taken at the time), and the administration decided to allocate only $1 billion to the VSE's initial implementation, hoping that an early end to the space shuttle and decisions by future presidents might make the program feasible. Despite this budgetary gap, NASA held a series of meetings with foreign governments to determine their possible interest, which led to a widespread exchange of views (including input from Chinese space officials) and the development of general planning documents, but fell short of outlining specific outside pledges of funding. For domestic reasons, NASA did not allow foreign space enterprises to bid for so-called critical path technologies in the VSE, such as the launch vehicle, astronaut capsule, or Moon lander. These U.S. technologies included the proposed Ares I and V rockets, the *Orion* capsule, and the *Altair* lunar module. What did eventually emerge, however, was a 2007 document called the Global Exploration Strategy (GES), a vision for enhanced international space cooperation in settling the Moon and exploring Mars. In addition, the GES meetings helped stimulate a parallel process in which scientists from many countries formed the International Lunar Network in order to develop compatible software formats and other technical protocols for sharing data from various national missions planned for the Moon over the next decade.

The Obama administration's concern over costs and what some saw as a questionable scientific rationale for a lunar outpost led it to cancel the U.S. Constellation program and with it the VSE in early 2010. The administration's aim was to allow less-expensive, commercial transportation services to provide access to low-Earth orbit, thus freeing up funds for science and exploration. But the redirection away from the Moon called into question the GES and other cooperative mechanisms based on a U.S. return to the Moon and establishment of a lunar infrastructure. As a result, international cooperation stalled, and NASA turned to robotic missions to study the Moon in the near term, with an eventual, more modest plan for a human mission to an asteroid.

After the last space shuttle flight in 2011, the U.S. commercial company Space Exploration Technologies (SpaceX) succeeded in developing the first U.S. post-shuttle transfer vehicle to the *ISS* in 2012. A parallell commercial effort by Orbital Sciences Corporation docked an unmanned *Cygnus* transfer vehicle to the *ISS* in 2013. But the U.S. Congress has resisted this redirection of NASA funds to the private sector, in large part because of the loss of jobs by established NASA centers and contractors. The result is that congressional funds now support efforts (at reduced levels) by private teams and traditional NASA-supervised contractors, slowing each effort.

Beyond the Moon, NASA has a number of missions—many of them cooperative—already in space for study of the Sun (*Rosetta*, with ESA), Mars (*Curiosity* rover, *Mars Science Lab* with ESA and Russia, and Mars Atmosphere and Volatile Evolution Mission), Jupiter (*Juno*), Saturn (*Cassini*, with seventeen countries), Pluto (*New Horizons*) and two asteroids (*Dawn*). A major forthcoming initiative is the U.S.-led *James Webb Telescope*, a joint effort among NASA, ESA, and the Canadian Space Agency. This cryogenically cooled telescope will have seven times the light-gathering capacity of the *Hubble Space Telescope*.[17] It will be placed at a gravitationally stable position behind Earth in relation to the Sun in order to shelter it from sunlight and improve its view on the rest of the universe. NASA also has many ongoing and planned missions conducting remote-sensing of Earth with various types of instruments for monitoring the atmosphere, climate, and a vast array of agricultural, aquatic, and geophysical features.

In long-term international engagements, the Obama administration's 2010 National Space Policy extended the U.S. presence on the *ISS* through at least 2020 and pledged to carry out an eventual human mission to Mars. The policy also called upon the U.S. government to:

> identify potential areas for international cooperation that may include, but are not limited to: space science; space exploration,

including human space flight activities; . . . Earth science and observation; environmental monitoring; satellite communications; . . . disaster mitigation and relief; search and rescue; use of space for maritime domain awareness; and long-term preservation of the space environment for human activity and use.[18]

All evidence, therefore, points to greater civil space cooperation, particularly as the technical contributions of international partners continue to grow. But questions still remain about whether or not NASA—given its declining budget—will be the leader.

The Naval War College space analyst Joan Johnson-Freese has outlined a cooperative plan for NASA that she calls the Space Exploration Partnership, an "inclusive" international strategy aimed at restoring U.S. space leadership in human spaceflight.[19] The concept would involve including U.S. *and* foreign technologies in building the future flight infrastructure and developing plans jointly through a "space summit to bring in potential partners early and give them . . . a voice."[20] But the current U.S. budgetary chaos, congressional mistrust about China's intentions, and lack of domestic consensus about the best path forward have thus far caused this sensible plan to remain on the back burner.

However, the fourteen original GES agencies have begun yearly meetings of the International Space Exploration Coordination Group to continue exchanging information on their plans. Maybe this could be a starting point for a new model.

European Space Agency (ESA)

ESA has emerged since its formation in 1973 as arguably the world's leading organization for civil space cooperation.[21] By its very nature, ESA is a collaborative structure, with twenty member countries (as of 2013) and a budget of $5.6 billion. It is a pioneer in regional space cooperation in a world where the norm in most regions—such as Asia—is mutual mistrust, nationalism, and competition. To pay for its activities, ESA operates on a two-tiered finance model. All members are assessed fixed yearly dues, linked to the size of their economies, to cover administrative ocosts as well as expenses for so-called mandatory space programs. A second set of contributions involves voluntary funding for special space missions—some of them large scale—for which the members receive contracts for their aerospace industries based on the level of their contributions (the concept of "just return"). This allows countries that want to spend more on space to do so and to develop specific national capabilities and areas of specialization.

Given its ballistic missile experience and its strong commitment to space activity, France quickly emerged in the 1970s as ESA's leader and main sponsor in the launch sector. Its success in developing the Ariane series of launchers has given ESA the independence it sought from the United States, after disputes regarding NASA's refusal to launch European commercial satellites in the late 1960s.[22] Ariane eventually became a leader in the global space launch market and a competitor in satellite communications. Meanwhile, Germany paid the lion's share of the manned science laboratory called *Spacelab*, which it began developing in 1973 for materials processing research in collaboration with the planned U.S. space shuttle.[23] Other countries, such as the United Kingdom, specialized in satellite construction. Over time, a variety of countries developed important core competencies that have allowed ESA to establish sophisticated space capabilities and emerge as one of the leading spacefaring organizations via a collaborative approach. ESA is a core member of the *ISS*, to which it has contributed the *Columbus* research module and the Automated Transfer Vehicle (ATV), one of only three cargo-delivery systems tested and ready to service the *ISS* after the end of U.S. space shuttle flight in 2011. ESA launched its first ATV, called *Jules Verne*, in 2008, and followed with the spacecraft *Johannes Kepler* in 2011, *Edoardo Amaldi* in 2012, and *Albert Einstein* in 2013.

ESA's experience in collaboration with outside partners, including NASA, Roscosmos, and China, among others, has been one of considerable success, but also periodic failure. Germans complained that NASA treated them in a "condescending" manner during the *Spacelab* negotiations, reportedly leaving them feeling like Germany was viewed as a "developing country."[24] The United States has also backed out of some important collaborative projects with ESA, such as the International Solar Polar Mission. This project was planned in the 1980s as a dual satellite mission to study the Sun's poles. But ESA had to change the project to a solo spacecraft mission after budgetary problems led to NASA's withdrawal. Similarly, NASA's recent financial woes have caused it to pull out of a dual rover and sample return mission, called ExoMars and planned for launch in the next five years. Such U.S. failures to follow through on commitments to ESA have led some Europeans to prefer partnerships with Russia and other potential collaborators, including China. But these partners have not been without their own reliability problems. China initially joined ESA's long-delayed Galileo satellite navigation system, but purportedly stole design information and used it to help design its own system instead, which now broadcasts on the same frequency claimed by ESA. China is no longer part of Galileo. For these and other reasons, in part related to common national security and economic interests among their member states, ESA and NASA continue to cooperate on the *ISS* and on many other projects. ESA, for example, will provide instruments and an Ariane launcher for NASA's *James Webb Telescope* later in this decade.

Russian Space Agency (Roscosmos)

Roscosmos emerged out of the rubble of the Soviet space program in the early 1990s as the inheritor of a financially bankrupt but technologically skilled set of space design and production enterprises. It also inherited a reliable stable of launch vehicles, an orbiting space station (*Mir*), and unparalleled experience in long-duration human spaceflight. Its space science missions had achieved a number of successes during the Cold War, but were overall less impressive than those of NASA. Regardless, the new Russian government could no longer afford to support its space agency. As a result, Roscosmos used the confusion, political freedom, and financial exigencies born of the Boris Yeltsin administration to fundamentally reposition itself within the international space arena through a series of contracts with NASA, private U.S. corporations, and other foreign space agencies and companies. Over the course of the 1990s, it shifted from being the most closed space program in the world to being the most accessible and commercial.

Its financial situation improved modestly under Vladimir Putin's first two terms as president (2000–2008) as oil revenues began to come in and enhanced tax collections replenished state coffers. Still, space science remained a poor stepchild of human spaceflight as Moscow focused on restoring former military satellite constellations for intelligence-gathering and navigation purposes.[25] In this environment, Russian scientists focused their attention on forming partnerships with foreign projects. Only a very few high-profile Russian missions went forward, one of which was the *Phobos-Grunt* spacecraft for Mars exploration, undertaken in cooperation with ESA. The mission plan included a lander for the Mars moon Phobos and a spacecraft to return the first Martian soil samples. However, technical problems with the computer system caused a failure early in its 2012 flight. It later turned out that Roscosmos had used electronic components from contractors that it had neglected to test adequately, causing in-flight rebooting at a critical time and then a communication failure.[26] The newly launched spacecraft was thus left stranded in low-Earth orbit without a command to fire the final thruster that would send it on its way toward Mars. It was a highly embarrassing and expensive mistake for Roscosmos. The Russian government was fortunate that the return of the large hydrazine-laden *Phobos-Grunt* spacecraft back into the atmosphere did not harm anyone, as it broke up and crashed into the ocean. Quality control remains a serious problem on Russian robotic missions, a symptom of Russia's lag behind other leading spacefaring nations in automated control systems and advanced microelectronics.

Where Russian science continues to benefit from the Soviet legacy is in its ability to provide critical, in-kind contributions—particularly in the form of launchers—that have enabled Russia to join a number of international space science projects without having to make equivalent financial commitments. An example is Roscosmos's decision in 2012 to participate in the former ESA-U.S. ExoMars mission.[27] Hoping to raise its science profile, gain access to useful data, and build ties to Europe, Roscosmos donated a Proton launch vehicle to deliver the first of two spacecraft in 2016 and will build a major portion of a second planned spacecraft scheduled for launch in 2018.

In human spaceflight, Russia has retained its position as the world leader through its robust launch systems and reliable access to the *ISS*. Russia's failure to meet a number of space station deadlines because of its budgetary problems in the 1990s initially caused NASA to question its decision to collaborate with its erstwhile adversary. But Roscosmos eventually provided critical modules and transportation of materiel and astronauts. This proved very fortunate for the United States. Twice—following the *Columbia* disaster in 2003, which led to a two-year shutdown in the U.S. space shuttle program, and now since the last shuttle flight in 2011—Russian Soyuz rockets have provided the sole means of access to the *ISS*. Fees paid by foreign governments for transportation to the *ISS* provide a significant portion of Roscosmos's budget. But such support will likely be jeopardized once the United States develops new commercial or civil transportation systems in a few years or if ESA or Japan decides to add the life-support equipment and reentry shielding needed to human-rate their respective ATV and HTV cargo vehicles.

Russia has an ambitious plan for modernizing and expanding its space science and human spaceflight infrastructure. Announced in the form of a legislative package introduced in 2012 to the State Duma for development of the space sector up to 2030, its goals include raising the number of space science missions from one a year to at least three per year, updating technology throughout the space complex, activating the long-planned Svobodniy launch facility in the Russian Far East (to end Russian dependence on the Baikonur site in now-independent Kazakhstan), and landing its first cosmonauts on the Moon by 2030.[28] The plan pledges increases in space funding in its initial years but indicates an expectation that some elements will become self-funding later on (although *how* that will be achieved is not fully explained). Given Russia's history of failing to meet space spending targets, it remains uncertain if Moscow will achieve its goals. But it is likely to remain a major participant in international cooperative projects due to its broad experience and capabilities.

Japan Aerospace Exploration Agency (JAXA)

Japan has long been the leader among Asia's space programs in the areas of space science and human exploration. It has carried out a number of world-class scientific missions that have made new discoveries and has had more astronauts in space than China (albeit via U.S. and Russian rockets). But the JAXA organization itself emerged only in 2003 in response to a perceived crisis in Japanese civil space activities: a series of embarrassing launch failures and ongoing bureaucratic infighting among the various organizations responsible for space technology. JAXA represented an effort to create a more unified organizational framework and to revitalize Japanese interest in space activity, particularly in the presence of perceived domestic technological malaise and emerging regional space challenges from China.[29] Through a series of successful launches of its H-IIA and larger H-IIB boosters and the conduct of several high-profile space science missions, JAXA has gotten the space program back on track.[30] The *Hayabusa* spacecraft became the first man-made object to land on an asteroid in 2005. Although the mission experienced some technical problems, JAXA managed to obtain particle samples from the asteroid and blasted off toward Earth in 2007, landing safely in Australia in 2010. Japan's *Kaguya* lunar mission conducted the first high-definition mapping of the Moon from late 2007 to June 2009, offering breathtakingly precise images and video on the JAXA website. JAXA also proved its "green" credentials by launching the *Greenhouse Gases Observing Satellite* in 2009. Finally, JAXA began ramping up its astronaut missions to the *ISS* and began delivering elements of its *Kibo* research module to the station in 2008, while conducting the first successful flight of its own space supply tug, the H-II Transfer Vehicle (HTV, or Kuonotori, in Japanese) to the *ISS* in September 2009. It carried out successful follow-on flights of the *HTV-2* and *HTV-3* spacecraft in 2011 and 2012. Japan has now entered rarefied company in joining the United States,

Russia, and ESA as the only countries qualified to service the space station, although not yet with passengers.

At the international level, besides its close cooperation with other members of the *ISS* team, Japan has sponsored the Asia-Pacific Regional Space Agency Forum (APRSAF) since 1993.[31] This body is a voluntary association of space research institutes and agencies that meets yearly and in ongoing working groups to share information on space scientific developments and missions. APRSAF also provides training and hardware for less-developed spacefaring nations in Asia. It is clear that while Japan is promoting scientific cooperation in Asia, it also has political motives. For example, Japan is providing China's rival Vietnam with more than $1 billion in assistance for the construction of a national space center and for the purchase of two Japanese synthetic-aperture radars for Earth remote sensing.

Japan's plans for the future include a Venus probe (*Akatsuki*) already en route to the planet, a second asteroid mission (*Hayabusa 2*), and the *BepiColumbo* dual spacecraft mission to Mercury being conducted in cooperation with ESA. JAXA has also committed itself to maintaining its role as one of the leading countries in studying greenhouse gases and climate change, which it sees as an important part of its unique scientific mission in space.

China

Critics argue that China has no "civil" space program because of the dominant role of the People's Liberation Army in its space activities. But, just like the Soviet Union during the Cold War, China conducts a wide range of space science and human exploration activities that have very limited or no military value. Instead, these missions are a means of expanding China's technological base, building domestic political support, and increasing international prestige. Military personnel play a central role in the country's launches and in its spacecraft operations; civilian scientists and technicians are engaged in space research and human spaceflight and cooperate with counterparts in many countries (although, at present, not with the United States, due mainly to opposition from the U.S. Congress over human rights issues and export control concerns).

China has a number of domestic space bodies that share responsibilities for different aspects of space science and human exploration. These include the China Aerospace Science and Technology Corporation (CASC), which encompasses launch-related enterprises, human spaceflight, and some 100,000 employees and the China Aerospace Science and Industry Corporation (CASIC), which conducts space and satellite research for military and civilian purposes and has enterprises numbering some 180,000 employees.[32] Although identified by the government as its civil space agency in 1993, the China National Space Administration (CNSA) remains a small organization focused mainly on international outreach, without any independent research, production, or operational facilities of its own.

China's progress in space science and human spaceflight in the years since 2003 has been impressive, although it has thus far largely been playing catch-up to the world's leaders.[33] Beginning with its first manned flight on *Shenzhou V* in 2003, China has built a steady record of accomplishment, sending two and three taikonauts into orbit at a time, conducting a spacewalk, and carrying out taikonaut visits to its orbiting *Tiangong 1* research module, a building block for a planned 60-ton space station. China is motivated by a desire for independent technological capability and by its exclusion from participation in the *ISS* because of opposition from the United States and Japan (a policy criticized by many U.S. scientists). With its growing economy and increasing space budget, China has moved forward on its own. Today, as the space analysts Handberg and Li argue, "China may not be in a direct race with the space pioneers but it is in a race with its own ambitions."[34]

In the field of space science, China's research program has grown steadily more sophisticated. While it still lacks experience in deep-space operations, it has continued to build its capabilities in near-Earth space. The *Chang'e 1* lunar probe conducted mapping operations from 2008 to 2009 as part of China's planning for follow-on lunar sample and automated rover missions.[35] The *Chang'e 2* mission, launched in 2010, conducted lunar mapping operations, then moved in 2011 beyond Earth's orbit, where it has tested some of the more sophisticated communications and controls that China will need to conduct future deep-space missions. The *Chang'e 3* probe will land on the Moon and conduct experiments.

To date, China has experienced a steady record of successes, while also benefiting from comparatively lower labor and hardware costs than other leading space programs (except India). Questions remain about how its space science and human spaceflight programs will fare in the future in case of operational problems, taikonaut deaths, budgetary difficulties, or problems moving beyond the limits of some of their foreign-derived technologies.

On the international front, China has been an active partner in both space imports and exports. It has sought to acquire space technology from leading companies in Europe and Russia through purchases and joint technology development programs. These efforts have included, for example, major purchases of human spaceflight technology and know-how from Roscosmos in the early 1990s and the formation of a joint company between Tsinghua University and Britain's Surrey Satellite Technology, Ltd., for work on small satellites.[36] In addition, China has purchased spacecraft control technology from Germany. But China has also conducted extensive outreach to less-developed countries through an organization it formed in 1992 called the Asia-Pacific Multilateral Cooperation in Space Technology and Applications (AP-MCSTA) and a formal, membership-only body called the Asia-Pacific Space Cooperation Organization (APSCO), formed in 2008 and modeled on ESA.[37] China has provided ground stations and access to satellite information to member states of APSCO, while also offering technical training. However, China's goal of heading an organization of peer space powers has thus far proved elusive, given the limited capabilities of current APSCO partner states: Bangladesh, Iran, Mongolia, Pakistan, Peru, Thailand, and Turkey.

Looking ahead, China shows no signs of slowing down, and indeed is increasing its infrastructure for expanded human spaceflight and deep-space missions. It will soon open a new launch facility on Hainan Island that will allow it to lift heavier payloads into orbit (although not yet equal to the U.S. space shuttle or Saturn V), as well as to access more favorable orbital inclinations closer to the equator. China's Long March 5 booster, expected in 2015, will be able to lift 25 metric tons into low-Earth orbit and 14 tons into geostationary orbit, while also possibly servicing deep-space missions. In addition, communications matter also. For this purpose, China is building two larger-diameter deep-space antennae, reportedly for future missions to Jupiter and the Sun. With its robust and growing infrastructure for space, China is well positioned to expand both its human spaceflight program and its space scientific and exploratory missions. As long as its economy continues to grow, analysts expect to see these components—which have brought China's government considerable domestic and international respect—develop accordingly, despite possible political obstacles and the challenges of U.S. reforming its state-controlled industries.[38] At some point in the not-too-distant future, China will likely join the ranks of the leading spacefaring countries in not just copying previous feats by others but also leading major, first-of-their-kind international scientific missions.

Indian Space Research Organization (ISRO)

India has conducted a range of space operations since its first satellite launch in 1980. But, in contrast to almost all other space programs, it traditionally neglected space science and shunned human spaceflight, except for a single cosmonaut mission aboard a Soviet spacecraft in 1984. Instead, its main focus was on providing practical space applications to its large and widely dispersed population. This tradition followed on the instructions of India's first prime minister, Jawaharlal Nehru, who called for science to serve the people and help lead the process of economic development of the country.[39]

But India's recently increased role in the international economy and its continued rivalry with China in both the military and the political spheres have resulted in a major shift in Indian space policy in the past decade. ISRO has now begun to undertake missions and adopt goals it once deemed wasteful and even frivolous: high-prestige space science and independent human spaceflight.[40]

Previously, India acquired space technology from both of the two superpowers but sought to follow a course of not becoming overly dependent on either. It cooperated sporadically with the United States in launch-related and Earth remote-sensing technologies, and with the Soviet Union in satellite design and solar cell technology. India also received assistance from Japan, West Germany, and France.[41] India's nuclear test in 1974, however, caused problems, in particular for its relationship with Washington, which opposed its nuclear and missile activities.

As India tried to develop the technology to launch larger rockets into space, it needed to master cryogenic engine technology. With the Soviet Union on its heels financially in the late 1980s, India sought to acquire production technology from the Soviet commercial space agency Glavcosmos. After the Soviet breakup in 1991, stiff U.S. sanctions on Russian enterprises and positive incentives through the promise of U.S. commercial cooperation led to Russia's agreement to provide cryogenic boosters, but reportedly without the associated production technology.[42]

India's series of nuclear tests in late May 1998 caused international opposition to new forms of technology sharing, particularly in the space-missile field. The United States sought to block India's further acquisition of technologies that could promote its nuclear weapons delivery systems. Strong U.S. sanctions banned scientific and technical cooperation. However, the September 11, 2001, attacks on the United States caused the George W. Bush administration to reevaluate its policies and to adopt instead a strategy of rapprochement based on common goals of democracy and anti-terrorism. As part of the Next Steps in Strategic Partnership agreement reached in 2004, the two sides began to put mechanisms in place for the loosening of U.S. export controls for the purpose of facilitating future cooperative missions in space science.[43]

China's first human spaceflight mission in 2003 provided India's science and technology structures with a rude shock. Faced with a swift and serious decline of prestige within Asia, ISRO recognized that the status quo in Indian space activities seemed woefully inadequate and risked further domestic and international degrading of its reputation, as well as broader damage to the country's position in international science and technology and political influence.

The first mission that emerged to meet this challenge was the *Chandrayaan 1* lunar orbiter in 2008. This project represented ISRO's first major scientific mission and included experiments and equipment from U.S. and German scientists. The politics of this move proved initially difficult, however. NASA required that ISRO sign a "technical assistance" agreement to govern its handling of the U.S. hardware and its non-acquisition of the technology.[44] ISRO officials viewed the requirement as an odious neocolonial vestige that smacked of the old U.S. denial strategy, and they initially refused to sign it. Only after NASA officials explained that all U.S. partners (including Russia) had signed such agreements and that the experiments could not be flown without the agreement did the Indian side finally agree to approve it. In the end, India's inexperience resulted in the mission's suffering a variety of technical problems, from overheating to attitude control, which eventually caused the spacecraft to crash into the lunar surface in the late summer of 2009. Little of this information reached ISRO's international partners in a timely manner, leading to complaints on the NASA side. Nevertheless, the mission could be reported internationally as a scientific success, thanks to the NASA equipment onboard, which determined the presence of abundant water ice on the Moon. Strong political pressure from Washington and New Delhi for enhanced bilateral cooperation suggests that, despite problems, the U.S.-Indian space relationship will continue to grow.

In seeking to move into human spaceflight, India has contracted with Russia since 2009 to purchase technology for life-support systems and thermal control elements. In addition, India has sent technicians to Russia and will eventually select and train an astronaut corps for its first independent human spaceflight (initially planned for 2016, but recently postponed until 2021). A preliminary flight aboard a Russian spacecraft will occur before that date as well. Nevertheless,

India's decision to undertake the significant costs and risks of developing a human spaceflight program indicates the political and even geostrategic pressure of civil space dynamics. India is also working to develop a lunar lander and rover and launched a spacecraft to Mars in November 2013, its first deep-space probe.[45] The Indian government recently expanded ISRO's budget by nearly 40 percent to handle these new commitments. To counteract domestic criticisms about the sudden expenses being taken on for high-prestige science and exploration, ISRO officials note the lower costs of its missions compared to those of other major spacefaring nations. Still, India clearly feels the need to step up its game in civil space in order not to fall further behind China.

TRENDS IN CIVIL SPACE COMPETITION AND COOPERATION

Despite the competitive motivations of many of the world's leading civil space programs, it is not a foregone conclusion that their respective activities will result in future conflicts. The international scientific community cooperates more closely than ever before because of the influence of the Internet and the relative ease of sharing data (and, indeed, the difficulty of controlling it). There is also pressure on all space programs to participate in missions that are going to *succeed*, which provides incentives to share both costs and technology, at least in terms of putting experiments and equipment on shared "buses." For the country building the main spacecraft, there is an incentive to reduce construction and launch costs by reaching out to others. The *ISS* is a good example. The spacecraft has cost much more than anyone anticipated, approximately $150 billion in total contributions pledged up to 2015 for what was once estimated to be a $30 billion project.[46] The initiator of the project, the United States, has shouldered about $126 billion (or 84 percent) of the total expense, although it has received valuable technology for the station from its partners. Of course, operating the station after 2015 will cost money too. For this reason, a number of *ISS* partner nations are considering opening the station to new members. In December 2011, ESA's director of human spaceflight and operations proposed that China become a partner in the *ISS*, saying that Chinese participation "offers great potential."[47] The United States has mentioned India as a possible member as well. But adjusting the existing membership structure and access agreements would be complicated. Still, with adequate political will, such an expansion of the station's membership is conceivable and might promote future lunar cooperation as well.

A critical question in the international civil spaceflight community is where to go next. Until the U.S. cancellation of its Constellation program, all of the major spacefaring nations had begun to line up to participate to a greater or lesser extent toward the Moon. NASA's decision in 2010 to shift to a possible asteroid mission threw these plans into confusion. As mentioned earlier, NASA had previously sponsored a series of meetings in the context of its lunar plans, one resulting in a cooperative Global Exploration Strategy. As the longtime space analyst John Logsdon observes, "The sudden U.S. shift away from the Moon as the initial destination for resumed exploration has undermined the momentum gained through that collective effort."[48] Moreover, compared to the possible benefits of lunar settlement, Logsdon argues regarding the list of most likely asteroids that "none of the candidate objects seems to justify more than a relatively brief, one-time visit."[49] A related problem is that other countries seem dubious of the goal, putting U.S. leadership into doubt. Finally, even recent studies indicate that the cost of visiting an asteroid may far exceed that for a return to the Moon because of the need for new technologies and the sheer distance from Earth of likely asteroid candidates.[50] Strong congressional criticism of funding even for studying the asteroid mission leaves the U.S. program in limbo. Regardless of what NASA chooses to do, it is likely that several countries will return to the Moon, as the nearest and arguably most logical destination to begin "settlement" of space. What country will lead them and who else will participate remain to be seen. It is still possible that U.S. legislators (who supported Constellation) or the activities of other countries may "drag" the United States back to the Moon. The other possibility for the United States might be a commercially led mission.

COOPERATIVE APPROACHES ON ORBITAL DEBRIS AND NEAR-EARTH OBJECTS?

As argued earlier, shared environmental challenges in space may have a restraining influence on more-hostile forms of national competition. Scientists are warning increasingly that our ability to conduct space activities, particularly in Earth orbit, is dependent on greater efforts to control and eventually reduce orbital space debris. A report issued in 2011 by the National Research Council argued that the amount of debris already in Earth orbit has reached the point at which even with no new launches, additional debris will be created in the coming years through collisions of existing debris, yielding more hard-to-track (and hard-to-avoid) hypervelocity hazards.[51] Tracking debris has always been a challenge, given the limited range and discrimination capabilities of ground-based radars.

Improving space debris management and space traffic control will require more and better radars in more dispersed locations. Linking this information into a useful network will also require operational cooperation among national space control centers. As U.S. deputy assistant secretary of state for space and defense policy Frank Rose stated at a meeting in Japan in February 2012, "International cooperation is . . . necessary to ensure that we have robust situational awareness of the space environment. No one nation has the resources or the geography necessary to precisely track every space object."[52]

For this reason, the United States has undertaken a series of discussions, beginning with its European allies and Australia, to expand its network of space-tracking facilities in the hope of increasing the current catalog of detectable orbital debris from around 17,000 objects[53] to around 100,000 or more in the coming decade through the use of more precise radars and optical telescopes. Besides the adoption of best practices for reducing the release of orbital debris in space operations, tracking and avoiding debris is the space community's best option for preventing collisions, such as the *Iridium 33-Cosmos 2251* crash in 2009. Fortunately, to date no catastrophic orbital debris has collided with human-occupied spacecraft.

However, as the debris field increases and the number of people in space grows, the prospect of such deadly collisions becomes more likely. NASA scientists worry that we have reached a tipping point at which debris mitigation efforts may no longer be effective because of the quantity of uncontrolled material already in orbit. These observations raise the issue of liability. Here, the guidelines are somewhat confusing. Written during the Cold War, the 1972 UN Convention on International Liability for Damage Caused by Space Objects indicates the responsibility of the launching state or the country that procured the launch. But today, more complex shared liability is often involved, due to the common practice of selling control of older satellites to a third country or even a multinational consortium, suggesting a more complicated legal process in deciding upon and awarding damages.[54]

After China's ASAT test released large amounts of debris in 2007, experts speculated that countries whose satellites might be damaged in the future by a collision with that debris could sue China. But questions arose as to who would provide the evidence. The only tracking system capable of providing such data is the U.S. Joint Space Operations Center, which might be accused by China of providing biased information. In cases of collisions with human-occupied spacecraft, of course, the consequences would be much more serious. But the specific rules, standards of evidence, and mechanisms for enforcement of liability cases remain unclear.[55]

Options for reducing debris are still experimental, and each approach has its downsides. Lasers could be used, for example, to ablate space debris and eventually slow its orbit enough to help drag it back into the atmosphere more quickly. Such lasers are not overly expensive to build. However, shooting a laser into space poses possible risks to spacecraft, and the experience gained could potentially contribute to a space weapon capability. Similarly, who would operate such as a system and how would they gain permission to "take down" debris belonging to other countries?

Another option for debris removal is to send a spacecraft into low-Earth orbit to capture and thereby remove especially large pieces of debris. The U.S. space shuttle showed that this could be done with whole satellites, which it sometimes brought to Earth for repair and then returned to orbit. The main problem with this concept is fuel. The space shuttle conducted such missions aiming at single, high-value objects. Trying to conduct *multiple* missions at different altitudes and in different orbital planes with no obvious economic reward not only would be expensive but also would quickly exhaust the fuel-carrying capacity of any currently designed spacecraft. The limited carrying capacity of any current spacecraft would be another problem, as the vehicle would fill up quickly. Even the removal of only 10 percent of the largest objects from the 17,000 pieces currently detectable would be a great deal of work, take a long time, and probably cost tens of billions of dollars. Russia recently proposed the novel idea for a long-duration, nuclear-powered spacecraft to conduct the mission of debris cleanup robotically. Setting aside cost and the problem that no country has yet built and tested a nuclear reactor for propulsion (as opposed to electricity generation) in space, there is the possibility that the reactor could malfunction and deorbit, sending radioactive material into Earth's atmosphere and possibly onto a populated area. Such concerns would likely make this spacecraft politically unacceptable to a significant number of nations.

A less drastic approach, but one that would require additional mission planning and costs for all users, is the use of space tethers. This concept involves the deployment of a cable (possibly several kilometers long) near the end of a satellite's service life. As Earth's magnetic field added an electronic charge to the cable, it would exert additional drag on the satellite and speed its process of reentering the atmosphere. Such measures could greatly reduce the number of dead satellites, especially those cluttering up highly trafficked regions of space. The cost of such systems is not especially high, but they would add weight and thereby additional expense to a spacecraft's mission. Moreover, some mechanism would be needed to persuade other operators to adopt these same measures.

Given the problems posed by each of the most likely potential solutions, the debris conundrum is likely to remain with us for some time to come. In the meantime, countries will have to continue to do the next best thing: try to reduce the release of orbital debris in the first place, while saving onboard fuel to allow end-of-mission transfers either up into super-GEO orbit or down into atmospheric reentry burn-up. Fortunately, this work is the focus of considerable attention. The United Nations Committee on the Peaceful Uses of Outer Space (COPUOS) has a multinational working group studying mechanisms for developing "best practices" for the long-term sustainability of the space environment, with a major emphasis on debris control. But new problems are on the horizon. These include the growing population of dead satellites in super-GEO orbit and the widely planned future use of tiny cubesats without fuel or maneuvering capability, which could clutter LEO orbits with large numbers of uncontrollable objects.

Near-Earth Objects

A major concern that has occupied increasing attention in recent years is the risk to our planet posed by so-called near-Earth objects (NEOs). Known historical collisions or explosions of meteorites, such as the Tunguska incident in Siberia in 1908 that leveled some 850 square miles of forest and the smaller 2013 event over Chelyabinsk, have caused scientists and laypeople alike to worry about future events that might harm Earth's population or its environment. Recent theories about the cause of past ice ages linked to impacts from space objects have only heightened these concerns.

In the past, a key problem was identifying asteroids that might be headed toward Earth. Today, with various ground and space telescopes and radars, much more has been learned, and there is a much more complete catalog of NEOs, though it is far from exhaustive. Although more than 7,000 large NEOs greater than 1 kilometer in size have been cataloged (of which only 20 percent appear to have some chance of hitting Earth), scientists estimate that this represents only about 1 percent of the total number of objects that might pose risks to Earth.[56] NASA Ames Research Center director Pete Worden says that large NEOs more than 50 meters in diameter with the potential to cause damage equivalent to a 9.2-megaton nuclear weapon occur about once every one hundred years.[57] If we discount the much-less-powerful meteor involved in the Chelyabinsk incident, we may still be due for another one.

NASA expenditures on so-called planetary defense have nearly quadrupled since 2010, but still amount to only about $20 million per year, mostly for cataloging and characterization. While this amount may seem too little for something of such global importance, it is not clear yet what we would do about it if we had more information. No international body has given the United States either responsibility or authority to deal with NEOs. Russia has expressed interest in the problem as well, although perhaps that interest results from its surplus nuclear weapons, which some Russian laboratories would like to apply to the problem. U.S. hydrogen-bomb scientist Edward Teller once declared similar intentions at a conference back in 1992, saying, "I would love to blow up [the asteroid] Ceres!"[58]

Options for defense against NEOs do include nuclear weapons, but some experts have argued that this would only create a huge cloud of equally dangerous smaller particles as well as electromagnetic pulse radiation harmful to spacecraft. For this reason, some scientists believe that diverting dangerous asteroids may be a better approach. Even very minor course corrections through the use of conventional propellants, if done early enough, could push an asteroid headed for Earth onto a safer course. This problem is one that is receiving growing attention as scientists—some backed by private money from Google, Facebook, and eBay entrepreneurs—develop a better catalog of threatening objects and begin to devise possible technologies for their deflection.[59] Planetary defense against NEOs qualifies as one of the most common interests among all countries on Earth. The question will be whether scientists and space agencies can develop a useful collaborative approach to build a NEO catalog, develop (and possibly test) defensive technologies, and then figure out a decision-making system for which ones should be used and when. Of course, liability for any mistakes could also become an obstacle, given the very serious implications of getting such actions wrong.

Besides the risk of a major asteroid or meteorite hitting Earth, so-called planetary protection is another concern. Since the first astronaut flights to the Moon, scientists have worried that contaminants from space might create incurable epidemics or other health or environmental problems on Earth. Film footage from the Apollo missions to the Moon routinely showed astronauts getting into special isolation units for testing and observation after their missions. What was less often shown was the ocean recovery of these same astronauts by U.S. Navy personnel, which clearly involved their contamination. Fortunately, the lunar environment, with its lack of an atmosphere, proved inhospitable to life and no harmful germs were transferred. Whether this will continue to be the case with sample return missions from Mars or other celestial bodies remains to be seen. So far, no life has been identified on any object reached by humans or by robotic missions in space.

The flip side of this problem is the related concern that visits from Earth probes and astronauts could somehow contaminate planets, moons, or asteroids that have environmental conditions reasonably similar to Earth's. Such contacts could conceivably alter the ecosystems of these planets by introducing life or by altering life forms not yet known to exist. As noted earlier, the 1967 Outer Space Treaty specifically requires that countries engage in space activities in a manner that will prevent the "harmful contamination" of celestial bodies (Article IX). As a result, the major spacefaring nations have frequently decontaminated their spacecraft by subjecting them to high heat or other processes intended to kill microorganisms that might have attached themselves to their spacecraft going to planets or moons with the possibility of sustaining life. For example, in 2003 NASA kept its *Galileo* spacecraft from crashing into the Jovian moon Europa—which is believed to have the possibility of harboring life—in order to avoid its possible contamination, sending it instead for certain destruction in Jupiter's inhospitable atmosphere.[60] It remains to be seen whether all countries will follow such practices as similar missions increase in number.

Scientists have also raised concerns about future pollution of pristine (albeit inert) environments in space, like the Moon's surface, if lunar mining operations dump toxic chemicals or other harmful substances or if repeated takeoffs and landings stir up regolith. Concerns have been raised especially about planets of high interest, like Mars, and a variety of moons around major planets that might have conditions conducive to life. In 1969, 1992, and 2002, the International Council of Scientific Unions' Committee on Space Research (COSPAR) issued guidelines on contamination and sterilization requirements to be followed by all scientific missions.[61] COSPAR categorizes risks according to five mission categories depending on the object's environment and the nature of the activity. Coordination among scientists and possibly the development of new rules and monitoring mechanisms by COSPAR or other bodies for both scientific and commercial entities may alleviate some of these concerns.[62]

CONCLUSION

A primary stimulus for U.S. and Soviet space science and exploration during the Cold War was competitive politics in a race for global supremacy that people on both sides generally supported. But this rivalry fueled mistrust and military tensions, as well as wasteful duplication of scientific effort. Today, U.S. and Russian civil space missions typically must be justified on the more tremulous ground of scientific value or service to national economic aims. But more amorphous goals such as international prestige have not disappeared entirely. In other countries, such as in today's Asian political scene, traditional competitive motivations still prevail frequently and help drive funding. Indeed, some leading spokespeople for increasing U.S. space science funding, such as the physicist Neil deGrasse Tyson, speak almost wistfully about past political hostilities and promote new competition (such as with China) to stimulate NASA spending.[63] But these arguments neglect the frequent accompaniments to such rivalries: increased military buildups and hair-trigger space tensions.

Instead, space politics and exploration in the twenty-first century might be better served by developing more innovative and mutually beneficial collaborative approaches to mission planning, funding, and operations, as has occurred with the *ISS*. Indeed, the ESA official Andreas Diekmann argues that the *ISS* should be viewed as a model for the future of space science.[64] But he contends that the membership must be broadened going forward. He proposes that the remaining years of the *ISS* might be a "test bed" for such a concept, which would presumably have to include both India and China—quite a tall order. Still, the alternative may be worse—either no funding for such missions or a bruising competition that is likely to result in military tensions and a related space arms buildup.

President Bush's second NASA administrator, Michael Griffin, supports such efforts. Over the hostile politics that have dominated the U.S. Congress, he has called for NASA cooperation with China in space science and in human spaceflight. While recognizing existing hurdles and differences in history and political culture, Griffin argues, "We should go that extra mile" to achieve scientific cooperation with China, given its potential benefits for reducing political tensions, stimulating joint research and missions, and making sensible use of the finite space resources available among spacefaring countries.[65]

Indeed, one incentive for such efforts is the high cost of major space science missions. The same incentives exist in major commercial space ventures, which have in the past two decades driven even erstwhile adversaries into highly integrated forms of self-interested cooperation. This topic and the challenges involved in carrying them out are the focus of chapter 4.

4

COMMERCIAL SPACE DEVELOPMENTS

Many readers were surprised in September 2012 when the Sunday Travel section of the *New York Times* devoted its whole front page to a story under the bold headline "Out of This World! Space, the Ultimate Getaway."[1] The author made the point that this was "not science fiction" any longer, given the progress made by Virgin Galactic, XCOR Aerospace, Blue Origin, and other commercial space companies toward making private human spaceflight a reality. While the article recognized that the first customers would have to have considerable spare cash, it explained that—unlike Russia's charge of $20 million and requirement of months of training for an orbital spaceflight—the price of getting into space will soon be as low as $95,000 for a private suborbital flight and will require only a few days of preparation. This might not yet be spaceflight for the masses, but it will mark a major step toward greater accessibility for those who are interested in venturing into space.

Beneath these headlines, other areas of space commerce have already become major moneymakers. In the fifty years since the launch of the first satellite to carry transatlantic television signals—*Telstar 1*—space applications have grown into a $290 billion yearly industry, increasing by double-digit rates despite recent global economic difficulties.[2] While potential tourist travel has grabbed recent media attention, the real money to be made in space is from transmitting *information*. Companies have thrived by sending valuable information around the globe from point to point in less than a second and gathering data in the form of visual images, heat emissions, and radar signatures that have value for a variety of industrial, agricultural, governmental, and ordinary civilian customers. One need only think of how the combination of U.S. Global Positioning System (GPS) satellites and high-resolution imaging satellites have revolutionized civilian navigation, allowing the creation of services like those of TerraServer, Keyhole, Inc., and more recently Google Earth, which have shrunk the planet in ways that were not even imaginable in the 1980s. With handheld and vehicle-mounted locational systems, space has facilitated the emergence of a whole new generation of young people who will never have to learn to read (or fold) a paper map.

Much of the infrastructure that supports and operates space commerce is invisible to public view. It is based on launches that take place at half a dozen sites around the world, satellites orbiting Earth at altitudes between 200 and 22,300 miles, ground stations that receive and transmit the data to normal fiber-optic cables, and dishes attached to people's roofs for direct reception. These various space-derived systems have emerged in fits and starts over the past several decades and now form a central component of our globalized economy. As numerous studies have shown, a cutoff of these services would have a serious and possibly catastrophic impact on the world economy. Thus all countries share an interest in making sure we maintain their continuity.

But revolutions in space commerce, such as the one we may be on the verge of today, have been overpredicted in the past. In the 1980s, entrepreneurs and space enthusiasts suggested that alloys, crystals, and pharmaceuticals developed in the near-vacuum conditions of space would be superior to those that could be fabricated on Earth and would quickly become commercially viable.[3] This did not occur, because of both the adequacy of Earth-grown materials and the extremely high cost (and corresponding low volume) of space manufacturing. Similarly, in the 1990s many investors believed that mobile communications based on fleets of low-Earth-orbit satellites would put a satellite phone in the hands of every consumer. That did not happen either. Although Motorola invested billions of dollars in its Iridium constellation of more than sixty satellites, it turned out that linking cell phones to cheap terrestrial towers and connecting them internationally via fiber-optic cables under the oceans proved much cheaper. Iridium went bankrupt and has emerged with a new business plan based on a more limited market of clients in the military, ocean shipping, emergency services, and ground transportation fields, where continuous service and the absence of dead zones is a higher priority than it is for most consumers. Looking ahead, it remains unclear if services like mobile satellite broadband, space solar power, and private human spaceflight will reach the point at which they can become viable businesses, or whether the expected profits will be consumed by costs or, alternatively, doomed by lack of demand. Finally, a related question is whether competition for profits will lead to international conflict, as has often occurred on past frontiers.

To understand current trends in space commerce, it is useful to review briefly how today's main space services emerged, as well as the international rules and organizations that have been developed to govern them. We will then survey the key elements of the rapidly growing commercial space sector—communications, remote sensing, and launch services—as well as two fields now emerging on the horizon: commercial human spaceflight and space solar power. Some people question whether or not this recent growth is sustainable and whether or not plans for the opening of additional activities—such as mining of asteroids and other celestial bodies—will accelerate these trends even more. For these reasons, part of this discussion must include an analysis of potential obstacles to the expansion of space commerce, including the crowding of the radio-frequency spectrum, the limited availability of slots in the geostationary orbital belt, problems posed by growing space traffic, and possible shortages in the availability of the highly trained scientists and engineers we will need to develop new space technologies.

A BRIEF HISTORY OF SPACE COMMERCE

Space commerce emerged initially in the early 1960s from a U.S.-government-sponsored effort.[4] The U.S. Congress passed the Communications Satellite Act in August 1962, which created a monopoly corporation (Comsat) to market international satellite communications for U.S. commercial spacecraft launched on NASA rockets. After the success of *Telstar 1*, the U.S. government negotiated with interested parties to create the International Telecommunications Satellite Consortium (Intelsat) in 1964, in which Comsat took a majority stake, thus giving leading U.S. satellite producers effective dominance. Comsat purchased privately made U.S. satellites, such as the geostationary *Early Bird I* spacecraft in 1965, and provided services internationally to Intelsat's initial membership of eighteen countries,[5] which built or purchased the appropriate ground transmitters and receiving stations. Within a decade, Intelsat's membership grew to eighty-six nations, and the consortium became the overwhelming leader in commercial satellite communications worldwide.

With its Cold War mission long ago fulfilled, Intelsat—with 150 member countries—became a private company in 2001. Now headquartered in Luxembourg, it remains a major provider of international telecommunications.[6] But it has a number of competitors, including satellite giants such as SES, Inmarsat, and Iridium (reconstituted in 2007 after its bankruptcy). Communications services remain the single most lucrative market segment in space, making up 38 percent of total space revenues at $111 billion in 2011.[7] Direct-to-home television service from geostationary orbit has emerged over the last decade to seize nearly 80 percent of these communications sector revenues.[8] Other services include satellite radio and fixed and mobile communications services from geostationary and low-Earth orbit.

Historically, the United States played a leading role in establishing a second major commercial space sector: remote sensing. NASA provided free or subsidized ground stations to many countries and the services of its Landsat satellites, first launched in 1972. Landsat's low-resolution imaging of optical and other Earth-emitted radiated energy from its multispectral scanner provided a valuable array of services to scientists, land-use planners, water management officials, and city managers, among others.[9] The U.S. government transferred Landsat from NASA to the National Oceans and Atmospheric Administration in 1979, which continued to manage its operations and data until 1985, when the Reagan administration engaged in a failed attempt to privatize its services. Eventually, Congress refunded Landsat and transferred effective responsibility to the U.S. Geological Survey in 1998, where it remains today.

More precise Earth imaging of greater use to the military (compared to Landsat) was originally highly classified. But the French government decided in the 1980s to create a new market in high-resolution images by establishing the Satellite Pour l'Observation de la Terre (SPOT) Image company in 1982, which launched its first satellite *SPOT-1* in 1986.[10] This service challenged the more developed space reconnaissance powers—the United States and the Soviet Union—to reconsider their classification limits on imagery by offering 10-meter resolution.[11] Eventually, under the reformist President Mikhail Gorbachev, the Soviet Union created Soyuzkarta to begin competing with SPOT Image with 5-meter imagery, dropping to 2-meter resolution by the early 1990s after the Soviet Union's breakup as the bankrupt Russian government struggled for cash. The United States finally responded under the George H. W. Bush administration by allowing U.S. companies to offer 3-meter resolution in January 1993. Today, companies like DigitalGlobe sell images to U.S. government agencies with a resolution as fine as 25 centimeters.[12] In fact, the commercial technology is now so good that U.S. providers need to dull the images in order to meet U.S. laws restricting the resolution of images for sale on the commercial and international market to 50-centimeter (half-meter) resolution.[13] These services have provided the processed imagery for the tremendously popular Google Maps service, which has revolutionized the global navigation and real estate fields, as well as many other business sectors that benefit from being able to analyze spatial relationships.

While some people (and government officials of would-be space powers) believe that commercial launch vehicles provide a major source of revenue, such devices actually represent a very small portion of the total commercial market in space. Of seventy-five space launches around the globe in 2012, for example, only twenty-three were commercially competed.[14] Historically, national governments have developed launch vehicles either for devoted space activities or as offshoots of ballistic missile programs, and they pay for these services at above-market rates to support domestic manufacturers. The Soviet Union/Russia and the United States began with national rocket programs to launch their satellites and still use them today. NASA also provided significant help to fledgling space science programs among its friends and allies during the Cold War by delivering these satellites to space free of charge. But NASA refused to provide free launches for commercial satellites because of its desire to support U.S. satellite companies working for the U.S. Comsat corporation, which eventually stimulated France and the European Space Agency to develop the Ariane series of rockets in the 1970s. This move helped establish the beginnings of the commercial satellite launch market. In the 1980s, the Reagan administration began to stimulate efforts to privatize American launch systems and make them available to the international market as well. Other commercial services emerged in China (Great Wall Industry Corporation) and later the Soviet Union/Russia (Glavcosmos).

However, the market for launch services has remained a highly protected one, with the vast majority of national launches remaining closed to commercial bidders. The United States has also typically required—even as it opened its commercial satellite market to international competitors—that emerging providers not so drastically undercut the prices of U.S. commercial launchers that they would put them out of business. For this reason, in return for the right to bid on launches of private U.S. satellites, China's Long March boosters and later Soviet/

Russian Proton and other boosters could be offered on only a limited basis, initially through a system of quotas and minimum pricing. Nevertheless, with its large stable of reliable, relatively cheap, and available launchers, Russia emerged quickly in the 1990s as the leader in the international commercial market. Besides France and China, only a few other countries have begun to enter this market—among them India, Japan, and Ukraine. More recently, small private U.S. enterprises—like the start-up companies Space Exploration Technologies (SpaceX) and XCOR Aerospace—have begun to offer services at competitive prices, benefiting from low overhead and highly motivated engineering staffs. This new wave of commercial competitors may well undercut government pricing models worldwide and shake up the international launch services market. SpaceX's Falcon 1e small satellite booster, for example, is being offered at the highly competitive price of $11 million per flight.[15] In the past, a U.S. manufacturer would have had to find a place on a larger and much more expensive booster to put up even a small payload.

One of the newest major sectors in the space commerce field is geolocation and navigation services, which make up 31 percent of the marketplace.[16] Information from GPS satellites serve the agricultural, aviation, shipping, trucking, and personal navigation sectors. It is worth remembering that this field did not exist at all before the United States orbited its GPS network (based on an earlier Navy system) and began providing its signals as a global public good free of charge. While this may seem like a giveaway by U.S. taxpayers to the rest of the world, U.S. companies have developed the bulk of GPS-related products and services and have reaped the associated revenues. Partly for this reason, many other countries are trying to develop their own precision navigation satellite constellations to stimulate their economies, although it remains unclear whether their products can replace GPS-based devices.

Compared to more-established commercial space sectors, human spaceflight has not historically been a profit-making enterprise. Instead, countries have sponsored their astronauts, cosmonauts, and taikonauts for purposes of national prestige and, to a lesser degree, scientific research. The United States and the Soviet Union also began to offer free flights to people from closely allied countries in the 1980s for political purposes. This model began to change in the late 1980s when the Soviet Union offered a six-month training course and launch services for $20 million to bring private individuals up to the *Mir* space station, which the increasingly impoverished Soviet space agency was having difficulty supporting. In 1990, the Japanese journalist Toyohiro Akiyama became the first fee-paying customer to visit space. The following year, Helen Sharman, a young Ph.D. chemist who had won a nationwide contest sponsored by a consortium of British companies, became the second space tourist to visit the *Mir* station. Russia later continued this practice by offering flights to sponsored or individually wealthy individuals. These services were also provided to foreign governments that wanted to send a scientist, engineer, or medical doctor into space. Largely for domestic political reasons and related goals of stimulating national interest in science, Malaysia took advantage of this opportunity in 2007 and South Korea in 2008. With the approaching advent of much cheaper suborbital flights and options for orbital flights on less-expensive, privately owned space stations, this sector stands to grow rapidly within the current decade.

Commercial Space Law

While few people are fond of regulations, space commerce could not have developed without some basic legal mechanisms to protect property and allocate limited resources in space among the competing international players. These rules and regulations emerged gradually, beginning in the early 1960s, in order to manage the growing space market and prevent possible disputes among its participants, particularly in regard to allocation of the radio-frequency spectrum and the limited stock of locations along the equator in geostationary orbit, the most favorable location for broadcasting and other communications satellites.

The International Telecommunications Union (ITU), formed in 1865 originally to govern transnational telegraph communications and later radio links, eventually took on both the responsibility of allocating satellite radio frequencies and that of parceling out the slots available above the equator in geostationary orbit. The ITU also helped extend principles established in other realms into regulations governing space communications, such as the right of non-interference with one's commercial signals. Other aspects of commercial space law stem from the 1963 UN space resolutions, the 1967 Outer Space Treaty, and the 1972 Liability Convention.

With the expansion of the number of space actors, especially in the past decade, the continued capacity of the ITU to handle challenges in space communications is being stressed and tested. As the editorial board of the industry newspaper *Space News* commented in a recent essay: "There was a time when access to orbital slots and broadcast frequencies concerned only the largest and most economically advanced countries. . . . But that era is coming to an end as the barriers to satellite-market entry come crashing down."[17] Today, virtually any country can purchase a satellite and pay a launch company to deliver it into orbit. As demand grows in Asia, Africa, and Latin America for the kind of reliable and abundant communications already available on other continents, the pressure on the existing commercial governance system is bound to become more intense.

The regulation of human spaceflight is still in its infancy compared to that of space communications. The 1968 Agreement on the Rescue and Return of Astronauts and Space Objects established a basic floor in terms of treating spacefarers as emissaries of all humankind and thus deserving of aid in case of distress, but it addressed mainly issues relevant to national human spaceflight programs rather than attempting to regulate private space tourism (which had not yet been envisaged). It also focused mainly on astronauts involved in orbital flights, rather than those that might take shorter, suborbital trips into space and return to Earth. As a result, a growing body of both national and international law is being developed today in anticipation of the emergence of regular commercial suborbital flights. These flights would likely involve airplane-like takeoffs, the firing of rockets to bring the vessel into space for a number of minutes so that passengers could experience weightlessness and see the curvature of Earth, and then return into the atmosphere and land. In the United States, the main concerns of the Federal Aviation Administration (FAA) have been to protect those people who choose to engage in suborbital (and orbital) flights using commercial services, but it also seeks to increase safeguards for those on the ground against hazards that such new services might create if these flights go awry.

At the national level, satellite launches and human spaceflight are affected by the fact that companies planning to put objects or people into space must obtain a license. In the United States, the FAA is the licensing authority, just as it is with commercial aircraft.[18] For satellite companies, this requires putting together a detailed flight plan from launch to the end of the satellite's service life, including (depending on its location) either deorbiting the satellite or boosting it into a super-geostationary orbit so that it is removed from the most heavily populated regions of space and does not collide with other objects. For human spaceflight providers, it will require flight plans more similar to aircraft flight routings, although some of the currently planned services will launch and return to the same location on Earth.

Another set of regulations that has traditionally affected space commerce is represented by national export controls laws. Such regulations are intended to control or prevent the spread of space technology. In fact, the United States and a number of other countries have controlled the transfer of space-related technology since 1957 in order to prevent foreign countries from acquiring possibly dangerous military secrets or uncompensated commercial benefits. The most serious concern has traditionally been the transfer of launch technologies that might be applied to a foreign ballistic missile program, particularly one that might deliver weapons of mass destruction against the United States. But various types of reconnaissance and communications technologies have also been the focus of concern. During the Cold War, such Western controls fell under the Coordinating Committee for Multilateral Export Controls (CoCom), which was formed in 1949 to prevent the outflow of technology to the Soviet Union. CoCom routinely denied any transfers of space technology. After the Soviet collapse in 1991, this body revised its regulations to allow commercial cooperation with Russia and other Soviet successor states. As part of this process, the Clinton administration removed commercial satellite technologies from the U.S. International Traffic in Arms Regulations (ITAR) munitions list in the mid-1990s to facilitate exports and the launch of U.S. commercial spacecraft aboard foreign boosters from both Russia and China, which offered U.S. satellite providers a means of significantly reducing overall flight costs and thus expanding their markets.

However, in the late 1990s, this issue became a point of contention when the U.S. Congress decided that China had gained space secrets of benefit to its nuclear delivery systems as a result of consultations between the U.S. companies Loral and Hughes and the China Great Wall Industry Corporation following accidents involving their satellites. Ironically, there were no charges of Chinese espionage or unauthorized access to the satellites themselves. While U.S. export officials were present at these meetings, the congressionally mandated Cox Committee report argued that secrets were still transferred through unsupervised conversations.[19] Some critics (including a number of technical experts) viewed the report as a poorly substantiated witch hunt,[20] but the result was that Republican-led legislators forced the Clinton administration to reclassify all space technologies (whether scientific, commercial, or military) as ITAR munitions items for the sake of U.S. export controls. This change in regulations caused commercial and civil space cooperation with China to grind to a halt. With other countries, it meant that U.S. companies or even federal agencies that wanted to cooperate with foreign space entities needed to prove that these actions (or even discussions) would not lead to the provision of militarily relevant know-how or technology, a requirement that created massive red tape and delays even when dealing with allies.

Space industry critics were vocal about the negative impact of these regulations on U.S. market share. Even run-of-the-mill equipment involving no sensitive information required special licensing and supervision after 1999. At a major space conference in February 2008 in Washington, D.C., involving a number of senior Bush administration officials, a lawyer from Bigelow Aerospace complained publicly that ITAR rules had required him to fly two U.S. export control officials at his expense so they could "babysit" a metal stand (which he compared to the legs of a table) used to mount a space hotel module it was testing in orbit.[21] Such ITAR horror stories raised costs for U.S. space companies and damaged U.S. exports.[22] Foreign countries—including allies like Canada, France, and Germany—reacted by limiting their purchases from, and cooperation with, the United States. The situation also stimulated foreign manufacturers to develop ITAR-free satellites in order to avoid the controls altogether. Finally, in December 2012, Congress adopted new legislation to reclassify commercial technologies in order to allow more normal U.S. space exports to all countries except China and North Korea, and any state sponsors of terrorism. But considerable damage had been done to the U.S. space industry.

TRENDS IN SPACE COMMERCE

The range of services available today from space is wide and growing. While governments drove space procurement and mission selection in the first fifty years of space activity, commercial dynamics are playing an increasing role in the overall market today. As technologies have moved from prototypes to commercial products and services, prices have decreased and satellite networks and support technologies have become part of the essential fabric of the global economy. A brief survey of trends in the leading sectors reveals a picture of a rapidly expanding array of products and services provided by a growing range of countries and companies, many of them multinational.

Communications Satellites

The international commercial satellite market continues to experience steady growth, as demand especially for direct broadcasting and mobile broadband services expands. The major shift over the past several decades has been the gradual decline of U.S. commercial hegemony. Whereas the United States held a dominant 73 percent share of satellite exports as late as 1995, a decade later it controlled only 25 percent of the market.[23] It has recently recovered to a 52 percent share. But market dynamics are now different. European and Asian producers (including Japan, India, South Korea, and China) have emerged as independent satellite manufacturers, no longer dependent on U.S. technology. Japan's Mitsubishi Electric Corporation, for example, has built satellites for Australia and Taiwan and recently scored a two-satellite deal with Turkey by bundling the sale with the Japan Aerospace Exploration Agency's provision of engineering and assembly training. China has used subsidized satellite exports to promote relations with political allies and potential energy providers, such as Nigeria and Venezuela. As one observer writes, "These nations don't pay in cash but in raw materials."[24] Recent progress by the Obama administration to shift commercial satellite exports to a new agency and cut licensing requirements has begun to redress the U.S. decline in market share.[25] Such U.S. manufacturers as Loral, Orbital Sciences Corporation, Boeing, and Lockheed Martin continue to set industry standards in terms of quality. Nevertheless, the growing competitiveness of foreign producers makes the likelihood of a U.S. return to its former dominance doubtful.

One of the fastest-emerging sectors in the commercial satellite realm is that of small satellites, or those generally under 500 kilograms in weight. Some of the most popular sizes are standardized cubesats (measuring 10 centimeters per side), other nonstandardized nanosats (weighing up to 20 kilograms), and tiny picosats (weighing up to 1 kilogram). The continuing progress of miniaturization in electronic components has allowed highly functional communications and remote-sensing satellites to be built in ever-smaller packages. Therefore, companies producing them are able to offer satellite services much more quickly and more cheaply than ever before, reducing costs from hundreds of millions of dollars often to a few million dollars or less. These payloads are also much easier to launch, and often fly at cut-rate prices as secondary payloads. Surrey Satellite Technology, Ltd., in the United Kingdom has emerged over the past twenty-five years as one of this market's leaders, selling its products to customers throughout Europe and in the United States, Kazakhstan, Nigeria, and China.[26] This part of the satellite market is bound to expand in the coming decade, making it possible for an increased breadth of users to afford ownership of devoted assets in space for commerce, research, and military purposes.

Remote Sensing

The field of space-based Earth observation is broadening in terms of both technology and customer base. Increasingly accurate optical-imaging satellites, hyperspectral sensors, and radar-derived products are becoming more widely available for agriculture, meteorology, disaster monitoring, forestry, military purposes, and traditional commercial uses, such as for real estate and urban planning. This diversity of services is expanding the pool of users and sharing costs of more expensive constellations and technologies. New countries purchasing and operating Earth observation systems for the first time will represent a third of all transactions by 2018, illustrating the dynamic growth of this sector.[27]

Climate change and natural disasters have been the focus of a variety of new Earth observation systems, including the Sentinel Asia program for early warning of tsunamis, hurricanes, and other dangerous natural events. Similar shared concerns in various other regions have promoted joint solutions and data sharing, as well as enhanced communications. Industry observers also believe that various imaging services will be able to continue to reduce their prices as the number of users expands, promoting the development of new products and data options. According to the Space Foundation, "These applications and services are poised to make Earth observation data increasingly integral to modern living."[28]

Launch Services

Although launching rockets is not as profitable as operating commercial spacecraft, it is a crucial and expanding sector that is gradually moving out from under subsidized and protected national programs. Such recent growth includes a range of new services, greatly increasing the options for companies, countries, universities, and other emerging clients. New launch locations are cropping up all over the world, as various countries and localities seek to capitalize on specific advantages, such as proximity to the equator (making it easier to reach geostationary orbit) or ability to launch over water or deserts (thus reducing risks to population centers). Among these new or proposed launch sites are equatorial Alcântara in Brazil, the desert in New Mexico, the Gulf of Mexico coastline in Texas, and Dutch Curaçao in the Caribbean. While not all of these sites are likely to become commercially viable, the rapid expansion of proposals in recent years shows rising expectations for new growth in the launch sector.

The dominant players in the launch business remain Russian companies, which have a long-established record of success and ability to provide relatively inexpensive and timely services with a relative minimum of red tape. With its stable of Proton, Soyuz, Dnepr, and Rockot launchers, some offered through joint ventures, Russia conducts more than double the commercial launches (ten in 2011) of the next-leading provider (France's Arianespace).[29] The prices of its various boosters are typically well below those of U.S. launch companies and those of Arianespace. As more competitors enter the market and prices drop further, subsidized national launch systems in various countries that benefited from closed markets for their governments' payloads will likely have even greater trouble competing if this bidding is opened up to foreign competition. Arianespace initially managed to succeed by establishing an efficient location for geostationary orbital launches along the equator in French Guiana. More recently, it has cooperated with Russian companies to expand its range of services, which now include the heavy-lift Ariane 5 ($220 million), the medium-launch Russian Soyuz, and the small- and medium-payload Vega ($24 million). Nevertheless, difficult market conditions forced ESA to provide Arianespace with a subsidy of 145 million euros (approximately $197 million) in 2011 to keep the company profitable.[30] Although occupying the high end of the rocket market, Japan's H-II A launcher has recently succeeded in attracting a limited number of commercial customers in Asia because of its high reliability and likely special deals for secondary payloads. The increasingly multinational nature of the space-launch field is also seen in the rockets used by the U.S.-based United Launch Alliance (ULA) and the Orbital Sciences Corporation. ULA's Atlas V employs the Russian-produced RD-180 engine. Orbital's Antares rocket—launched from sites in Wallops Island, Virginia; Cape Canaveral, Florida; Vandenberg Air Force Base, California; and Kodiak, Alaska—uses U.S.-made Aerojet engines but relies on Ukrainian fuel tanks.

SpaceX, based in Hawthorne, California, is one of the most promising newcomers to the U.S. launch market.[31] After several initial launch failures, this start-up company funded by the Internet entrepreneur Elon Musk has established the small-payload Falcon 1 and heavy-lift Falcon 9 as serious price competitors of traditional launchers. The company's workforce has tripled since 2011, numbering 3,500 employees by mid-2013.[32] Although benefiting from initial NASA and Department of Defense (DoD) contracts, SpaceX has succeeded in achieving its goal of entering the international market as well. Included in its backlog of $3 billion in orders are contracts with Argentina, Mexico, Taiwan, and Thailand.[33] Its streamlined manufacturing processes have allowed it to offer prices often 50 percent below those of Russian, U.S., and French launch companies. In a move unprecedented in the aerospace industry, Musk hired away the assembly-line manager from BMW's MINI Cooper automobile factory to help bring economies of scale to his facility.[34] The company's *Dragon* capsule was the first commercial spacecraft certified to bring supplies to the *ISS*. Since the capsule is pressurized and returns into the atmosphere for a soft landing in the ocean, SpaceX is working to human-rate a specially equipped *DragonRider* capsule with seating for seven astronauts. This will allow the company to bring crew members to the *ISS* as well carry space tourists. Its planned Falcon heavy-lift booster will offer 53 metric tons to low-Earth orbit, or roughly double the capacity of the former largest-delivery system, the U.S. space shuttle, for an estimated $80–$125 million (compared to about $350 million for NASA's former space shuttles).[35]

The lowest-cost niche in the current market is occupied by China's Long March family of boosters marketed by the Great Wall Industry Corporation and India's Polar Satellite Launch Vehicle (PSLV) and Geosynchronous Satellite Launch Vehicle (GSLV) marketed by the Antrix Corporation.[36] The increasing reliability of these services and their price competitiveness make them the go-to launchers for many newly

TABLE 4.1: Selected Commercial Launch Vehicles

Vehicle (Country)	Payload (to-low-Earth orbit) (lbs.)	Thrust (lbs.)	Rocket height (ft.)	Fuel	First launch
Falcon Heavy (U.S.)	117,000	4.1 million	224	Liquid	2014*
Long March 5 (China)	55,000	2.4 million	197	Liquid	2015*
Delta IV Heavy (U.S.)	50,000	1.7 million	235	Solid and liquid	2004
Atlas V (U.S./Russia)	45,000	1.9 million	191	Solid and liquid	2002
Proton (Russia)	45,000	1.5 million	174	Liquid	1967
Ariane 5 (France)	38,000	1.4 million	174	Solid and liquid	1996
H-IIB Heavy (Japan)	36,400	1.4 million	186	Solid and liquid	2009
Long March 3B (China)	30,000	1.7 million	200	Liquid	1996
Falcon 9 (U.S.)	29,000	1.5 million	224	Liquid	2010
PSLV (India)	8,150	1.1 million	129	Solid and liquid	2009

* Estimated

Sources: Space Foundation, "Space Infrastructure," *The Space Report: 2012* (Colorado Springs: Space Foundation, 2012), 70–79; "H-IIB Launch Vehicle," http://www.jaxa.jp/projects/rockets/h2b/index_e.html; Asif A. Siddiqi, *Sputnik and the Soviet Space Challenge* (Gainesville: University of Florida Press, 2003), 131; Space Exploration Technologies Corporation (SpaceX), "Falcon Heavy: Overview" and "Falcon 9: Overview," http://www.spacex.com/; Great Wall Industry Corporation, "LM-3B," http://www.cgwic.com/LaunchServices/LaunchVehicle/LM3B.html; Indian Space Research Organisation, "PSLV-C8" (product folder) (April 2007); Craig Covault, "First Look: China's Big New Rockets," Americaspace.org website, July 18, 2012.

emerging spacefaring nations. China is likely to increase its market share as more countries develop ITAR-free satellites and remove themselves from the U.S. ban on using Chinese commercial launches. The planned heavy-lift vehicle Long March 5 and the opening of its Wenchang launch facility on Hainan Island may also make China more competitive in servicing the geostationary orbital belt. China has strong domestic demand for its boosters, suggesting that its launch rate is likely to soon surpass even its high of nineteen launches in 2011, beating the United States for the first time. While this may be a point of concern for the United States, China is still playing catch-up as it seeks to populate its Beidou global positioning constellation and other satellite systems that the United States already has in operation. India is also likely to increase its commercial presence in the coming decade, largely due to price competitiveness. Singapore, for example, turned to India to launch its first domestically built satellite in 2011, and past PSLV clients include Argentina, Belgium, Germany, Indonesia, and South Korea. Other countries seeking to enter this end of the market include Brazil, Indonesia, and South Korea, only the last of which has thus far launched a satellite successfully.

Commercial Human Spaceflight

Perhaps the most exciting portion of the emerging commercial space market is the wide array of companies devising independent means of making space travel a reality for ordinary people. Space tourism is about where human air travel was in the 1920s. The concept has already been proven, but services for the average citizen have not yet emerged. For much of the rest of the decade, the cost of spaceflight will remain well beyond all but the highest fraction of upper-echelon earners. With suborbital Virgin Galactic flights scheduled to begin in 2014 at a cost of $250,000 per person, a mass market will not emerge. Nevertheless, this is a start, and experts believe the industry has reached a turning point and will soon begin to flourish.[37]

Part of the challenge facing reasonably priced space tourism relates to technology. It must be safe, it must be easily replicable, and it must not require the months of training currently needed for most orbital spaceflights. The *SpaceShipTwo* vessel, with up to six passengers and a crew of two, which was developed by Burt Rutan's Scaled Composites company for Virgin Galactic, launches aboard a conventionally powered aircraft (called the *WhiteKnightTwo*) from a runway and then separates and fires a rocket engine at high altitude that propels it just beyond the atmosphere for several minutes of weightlessness. It glides back to the same runway. Virgin Galactic's commercial service will initially operate from the Spaceport America near Las Cruces, New Mexico. The company won a major legal victory in 2012 when the U.S. government ruled that it could fly foreign citizens without the necessity of acquiring an export license.[38] This ruling gives Virgin Galactic—and future U.S. competitors—a much better chance of establishing a viable business. One of these, XCOR Aerospace, plans to offer initial service from California's Mojave Desert and possibly Curaçao in the Caribbean with the *Lynx* spacecraft, a small winged vehicle that launches from a runway and can take one passenger (plus a pilot) into the lower reaches of space.

The orbital spaceflight market includes several companies, all of which have had greater trouble moving their technology and business model forward toward viability. Bigelow Industries has launched NASA-derived test modules but has suffered from gaps in financing. Orbital costs are much higher for launch and maintenance, thus raising ticket costs and limiting the customer base. Russia currently charges NASA more than $60 million per astronaut for delivery to the *ISS*. Commercial services, including SpaceX's planned *DragonRider* capsule, Boeing's *CST-100* spacecraft, and Sierra Nevada Corporation's *Dream Chaser* vessel (all recipients of NASA seed-money awards), could put sharp downward pressure on pricing in this market. But as of this writing, such services are still a few years off.

Despite these challenges, commercial human spaceflight will likely revolutionize how people get into space by 2020. Prospective customers will no longer have to rely on government-controlled systems owned by Russia, China, or the United States. As a result, people will begin to think differently about space as the cumulative number of astronauts jumps from the current level of five hundred or so to the low thousands. The presence of ordinary citizens in low-Earth orbit will also put a much higher priority on keeping this area of space free of debris and other hazards. The concept of "safe access to space" will become increasingly important and will be driven not by scientists and the intelligence community, but by the growing population of space tourists, supported by the commercial space sector.

Space Mining

On the commercial space horizon, much interest is being paid to the prospects of mining and resource development on celestial bodies, including planets, moons, and asteroids. While the costs of production and transportation are likely to be very high, at least a few countries and private entrepreneurs believe that mining the celestial bodies (which contain materials not found on Earth) could offer the potential of great benefits for building structures in space, developing fuels for travel, or even returning particularly valuable minerals to Earth. The unique location of the Moon near Earth and the presence of rare helium-3 in its soil have stimulated interest in lunar mining. The reason is that the rare helium-3 isotope (blown to the Moon from the Sun by solar winds) has the potential to generate a nearly limitless supply of nuclear energy. But technical and cost challenges exist in separating it efficiently from the Moon's regolith, transporting it to Earth, and developing new nuclear reactors that can use it as fuel.[39] Still, former astronaut Harrison Schmitt and, separately, the Chinese government are among those that have stated an interest in its development. The real questions are who will actually invest the large sums of money needed to find out whether this resource is commercially viable and when. According to other analysts, radioactive helium-3 should instead be tapped in small quantities as a natural fuel for lighting on the Moon, rather than "strip mining" the Moon for sending large quantities of the material to Earth.[40]

The space scientist Paul Spudis predicts that regolith could be processed into blocks or other shapes useful for constructing living quarters, industrial facilities, or launch pads.[41] With recent indications of the presence of water on the Moon, other options may become available as well. As Spudis says, "Water can be processed into liquid hydrogen and oxygen propellant to re-fuel spacecraft both on the lunar surface and for export to cislunar space."[42] One expert in mining robotics, Greg Baiden, has outlined a complete design for a large-scale underground operation on the Moon to process and store significant amounts of these materials "in a manner that minimizes the amount of mining machinery and personnel required."[43]

Even more adventurist entrepreneurs are considering the mining of materials on asteroids within the solar system. Billionaire Google executives Larry Page and Eric Schmidt have joined film director James Cameron in backing an initiative known as Planetary Resources, Inc. Co-chaired by the space visionary and X Prize leader Peter Diamandis, this enterprise plans to conduct long-duration missions to asteroids to carry out in situ mining and return missions to Earth.[44] While critics doubt the economic viability of the company because of the currently high transportation costs, the concept could revolutionize how people conceive of resource availability, given the finite nature of Earth's mineral supplies.[45]

Space Solar Energy

With energy needs on Earth continuing to expand and fossil fuel resources showing signs of waning (as well as producing negative environmental effects), the prospect of harvesting solar energy from space presents an enticing option. While solar power collection on Earth has expanded rapidly in the past decade, it remains a tiny portion of overall energy used for electricity and other purposes. Much of the Sun's light energy is absorbed by the atmosphere and is limited by factors such as cloud cover and the unavoidable occurrence of night. Collectors based in space, by contrast, would have the considerable benefits of a much more powerful light source and one that could be collected twenty-four hours a day. Moreover, the Sun's energy is both nearly limitless and free.

The challenges to the successful commercialization of space solar power are threefold: the high cost of launching construction materials, the engineering challenges of building large collectors in space (ideally many times larger than the *ISS*), and the development of a safe and reliable transmission system.[46] Each of these problems is significant today, although progress is being made. Extremely light, carbon-based materials may be one means of reducing launch-weight demands. Commercial launch vehicles now coming online promise to reduce launch costs by at least half. Finally, experiments on Earth have shown some advances in beaming solar power from a receiving platform to a collector over a limited distance. But much more work needs to be done before even a prototype system is ready for space-to-space use, much less safe enough to beam energy to points on Earth. Still, the concept is a relatively simple one (compared to helium-3) and could have a significant effect on the energy field sometime in the next several decades, perhaps making the use of fossil fuels seem quaint.

PROBLEMS, OBSTACLES, AND DISPUTES

As noted earlier, two main dilemmas are putting past mechanisms of commercial space governance at risk in the early twenty-first century: more actors and fewer available resources in critical regions and radio spectra in near-Earth space. Recent disputes suggest that the existing regime is fraying at the seams and that conditions may worsen before they get better. Enforcement mechanisms are weak, and incentives to cheat on existing norms are relatively high, given increasing demands for access and perceptions of inequality by latecomers to space. Since power in space will undoubtedly become more diffuse as technology spreads, it is important to analyze these and other commercial disputes, which could spill over into conflict if not managed successfully.

Geostationary Orbital Slots

A key finite resource in space is the number of slots available in geostationary orbit above the equator in areas that provide coverage over the most populous regions of the world. These orbits are important for a range of tasks that require steady "staring" at the same points on the ground for communications, broadcasting, or early warning of missile launches. Problems have arisen in areas of the world where newcomers with limited funds or technology are seeking to acquire new slots or maintain a hold on the few allotted to them (requiring an active broadcasting satellite on station). One recent dispute involved the slot located at 34 degrees east above the Middle East. Iran possessed a claim but failed multiple deadlines for placing a satellite in the position and actually using it.[47] After the ITU had decided finally to remove Iran's unexecuted claim to the slot, the majority of countries at the World Radiocommunication Conference voted to restore Iran's right to the position after a slight compromise, setting a troublesome precedent.

A similar problem occurred recently with a slot claimed by Pakistan. After Islamabad had failed to occupy three of four slots initially allocated to it, the government finally managed to purchase a Hughes-built satellite from Indonesia already in orbit (and serving Turkey under a lease arrangement) and move it into the position.[48] Although the satellite had solar cell problems that enable it to provide only partial service, the move allowed Pakistan to retain its right to the slot until it can lease, purchase, or develop a devoted satellite for the position.

Jamming and Radio-Frequency Interference

A serious emerging problem is the increasing occurrence of continued electronic jamming of commercial satellite signals. In the main, this problem relates to authoritarian regimes seeking to keep information from entering their airwaves and being distributed to their populations. The main perpetrators of this violation of international telecommunications law on non-interference are isolated countries like Iran and North Korea, but other nations—such as Thailand—have resorted to jamming in times of national crisis. A particularly interesting and politically complicated case involved jamming, originating from Cuba, of U.S. uplink signals to the *Telstar 12* satellite in 2003. This satellite had recently begun broadcasting a U.S. Voice of America service in Persian that was critical of the Iranian government. After inquiries with the Cuban government, the interference was traced to equipment on the roof of the Iranian embassy in Havana, and Cuban officials shut it down.[49] But Iran continues to jam downlinks from facilities on its own soil of various Western satellites beaming U.S. and British Broadcasting Company programs into Iran.

A related problem is the increasing demand on the finite radio-frequency spectrum. Put simply, there are too many users chasing popular bandwidths, such as the C-band spectrum used both by many satellite communications providers for television signals and by Internet services for tower-delivered broadband service. Similar disputes have arisen surrounding a company (LightSquared) that in 2010 sought to acquire a license to offer mobile broadband services via satellite.[50] The problem was that its proposed L-band signal (from low-Earth-orbiting satellites and ground-based towers) would have drowned out weaker but arguably more important L-band GPS signals coming from medium-Earth orbit. U.S. regulators blocked LightSquared from acquiring a license to broadcast. Decisions by other providers and by other national regulators might differ for other parts of the globe, but the problem is not likely to go away.

A possible technical solution is the use of laser pulse communications—that is, the use of directed energy to carry large amounts of data. Such systems could greatly reduce the need for use of the radio waves in the first place. To date, the United States, France, and Germany have conducted experiments in this technology, whose part of the electromagnetic spectrum is not restricted.[51] But it requires highly accurate beams that must travel over long distances, as well as power sources adequate to deliver a strong enough pulse. Both tasks are difficult. There are also possible safety concerns in beaming data to the ground. Recent NASA and ESA investments in laser technology terminals, however, suggest that a small but growing share of communications will be transmitted through this means within this decade.

Space Traffic Control

If current trends continue, it is likely that today's approximately 1,050 operating spacecraft could double or triple by 2020, as more and more countries, companies, and other actors choose to invest in space-derived information and services. Much of this increased traffic will take place in low-Earth orbit, a region already beset with problems of identifying, tracking, and avoiding high-speed orbital debris. In fact, multiple levels of activities will have to be coordinated if this threat to space commerce is to be overcome successfully.[52] First, space operators will have to agree to provide more data about their activities in space. Space-launching countries are currently responsible for providing basic information about their spacecraft under the terms of the UN Registration Convention. But, as noted earlier, compliance has been imperfect and the reporting often late. Successful future traffic management will require not only advance information but also updated data on satellite maneuvers and orbital changes. While a small number of countries may try to assert national security prerogatives, at least getting the vast bulk of space traffic under better control will promote improved safety, which will be essential to space commerce. Technically, improvements will also need to be made in U.S. and international monitoring of space objects. This process is beginning with new space situational awareness collaboration and the expansion of the U.S. space-oriented radar system.

Private industry has also responded to the debris and traffic management problem. After being frustrated by attempts to acquire more accurate space traffic information about geostationary orbital satellites from the U.S. Joint Space Operations Center, three leading telecommunications providers (Intelsat, Inmarsat, and SES) banded together to form the Space Data Association (SDA).[53] Now joined by a host of other companies and some governmental organizations, the SDA has created a geostationary orbital belt database of satellite locations and provides information to its members in order to prevent collisions.

But new technologies and inadequate national legislation in the area of small satellites may confound efforts at traffic control in space over the next decade. Currently, a number of countries do not require licenses for the operation of cubesats, which they do not consider functioning spacecraft because of their lack of a maneuvering capability. As of 2012, only a few dozen cubesats had been placed in orbit. But with ESA plans to launch fifty cubesats in 2014 and with hundreds of universities and companies interested in using these cheap space vehicles, the prospects of collisions could increase dramatically. Moreover, since these satellites lack fuel, they do not have the capability of pushing themselves down to burn up in the atmosphere, meaning that they will remain "unguided missiles" until their orbits decay—often for a decade or more. In light of these problems, there are a number of proposals under consideration to require their registration and also to devise some means of speeding their orbital decay. Unfortunately, given the high demand for cubesats, the risk they pose to space traffic management is likely to get worse before it gets better.

Space Workforce Issues

Another key concern for the United States and Russia, in particular, is how to maintain and develop their valuable cadres of space engineers and scientists. With declining space budgets and limited plans for new missions, NASA and Roscosmos—the two leaders during the Cold War—now face an aging workforce and challenges for hiring promising young people for careers in space. The low salaries in government agencies are a particular problem, as are demographics, with a declining share of young people studying the requisite science and math courses needed to enter the aerospace field. The U.S. aerospace workforce declined by 3 percent in 2011, from 260,000 to 253,000.[54] Moreover, 88 percent of the workers in the field are older than 35. Some sectors of private industry, however, maintain their allure, especially in companies that are developing new means of getting into orbit, such as SpaceX and Scaled Composites/Virgin Galactic, and require fewer workplace restrictions than government-run programs do. Nevertheless, the U.S. military space sector bucked the overall trend and grew by 17,000 workers in 2011, due to steady government demand. Russian dynamics, based on more anecdotal evidence and budget trends leaning toward military growth, appear to be similar. Still, the trends pose challenges if the two countries hope to maintain their leadership roles. In Russia, the main emphasis remains on enterprises linked to Roscosmos, which explains why President Putin is pledging a raft of new missions and increased funding.[55] But whether his commitments will be met and whether prices for energy exports (on which the Russian economy depends) will remain high enough to fund this new activity remain to be seen.

China, by contrast, has a young and growing workforce in the space sector. As it expands its infrastructure for civil, commercial, and military activities and continues to increase its yearly number of launches, these trends are bound to continue into at least the near future. Estimates vary, but it appears that China has already surpassed both Russia and the United States in numerical terms, with more than 300,000 people in the space sector. China has also been more effective in pushing students into studying science and engineering. Where the country currently lags is in innovation, a critical component of becoming a leading spacefaring nation. Indeed, some academic research argues that while China's educational system excels in churning out qualified technicians, it is much poorer at fostering the development of innovative thinkers, commercial entrepreneurs, and technical risk-takers.[56] Given the heavy weight of state-run enterprises in China's space program today, these factors could hamstring China's future space program, particularly if the commercial space sector in the United States and other capitalist countries develops valuable new directions in space technology.

For these reasons, as the space workforce becomes more mobile and transnational, concerns about possible foreign espionage in the U.S. commercial space sector have increased. A number of recent cases involving theft or proposed sales of technology to China by corrupt industry personnel suggest that the issue is a serious concern.[57] The Federal Bureau of Investigation has prosecuted a number of individuals, both current and former Chinese nationals, who were caught attempting to transfer or sell space technology or design documents to the Chinese government or its enterprises. In one case, a former longtime Boeing employee and former Chinese national had removed copies of 300,000 pages of sensitive documents on various rocket systems and had been mailing them to China or transferring them through a Chinese consulate. In another case, a former Chinese national who was working for a U.S. satellite company had downloaded hundreds of sensitive technical files on a computer that he took with him to China. One Chinese national was caught attempting to export controlled U.S. computer chips made in Colorado to a Chinese space enterprise via Mexico by disguising them in packages marked as baby formula. Another former Chinese national was caught trying to purchase hundreds of radiation-hardened microprocessors used in space technology for transfer to China. These cases will require continued attention by the U.S. government and industry, as well as Chinese authorities, to halt these practices if China wants to resume commercial cooperation with the United States.

CONCLUSION

After fits and starts, the development of the commercial space industry seems to have hit its stride. Promises in the 1990s of a vast expansion in satellite services did not materialize due to the fiber-optic cable revolution. But it seems now that a critical mass of new technology, innovative services, and expanded space- and ground-based infrastructure has created conditions in which commercial space will flourish. While a sharp global downturn could reverse these trends, given the high costs of space systems, the steady growth of the space industry even during the 2008–2009 worldwide recession and its aftermath suggests that space services are now deemed essential and their expansion a necessity, not a luxury any longer. Civil space and scientific activities, by contrast, have not fared as well. As a recent industry-wide review summarizes, "The space industry has matured, [and] space products and services have become an integral part of daily life."[58]

But the increasing number of actors and emerging conflicts between the interests of long-standing space powers and those of rising newcomers in the commercial sector pose a number of challenges. Compliance with regulations and structures formed under different conditions during the Cold War is becoming a more serious problem than ever before. As *Space News* commented in its 2012 editorial on the tenor of recent discussions at the World Radiocommunication Conference, newcomers to the satellite communications market perceive a bias in the original allocation of geostationary orbital slots and radio frequencies. While they have had to go along with the leading spacefaring nations in the past, that could begin to change. These operators, *Space News* warns, "could . . . upend the entire regime—developing nations far outnumber developed ones in the WRC's one-country, one-vote decision-making process."[59] This situation poses a long-term dilemma for space governance prospects, at least in the commercial sector.

Looking ahead, the question seems less one of whether the commercial space field will grow than one of *how* it will grow and with what degree of stability. The one shared interest, however, among all commercial space entities is that the space environment remain one where conflicts can be settled short of war and that hazards such as orbital debris are kept to a manageable level. The vast majority of companies would also like to see the steady development and application of legal protections to space so that investments are protected and liability claims remain respected and enforceable. Unfortunately, the future is murky on these issues. Growing military and political tensions among a number of spacefaring countries make it difficult to confidently predict peace and stability over the next several decades. As in other areas, much will depend on the ability of the actors to reach beyond their individual interests to ensure the protection of their greater common interests. Growing collaboration across national borders in space commerce may make this more likely. Similarly, the sheer importance of commercial space services to the international economy should promote caution in putting any of these systems at risk, since the consequences of a long-term shutdown could quickly escalate into the billions or even trillions of dollars. Hopefully, this shared interest in maintaining favorable conditions for the continuation and steady growth of space commerce will urge restraint on those actors that might seek to disrupt the current system.

Chapter 5 examines military-related challenges and opportunities in space and seeks to develop an overall perspective on trends and possible outcomes over the next few decades. As in any field, different directions are possible, but each has certain advantages and disadvantages. Trade-offs are part of moving forward in any area. The question for the still-emerging field of space activity is whether the relatively peaceful international development of this environment is likely to continue or to be disrupted by the failure of self-restraint, dangerous new technologies, or the breakup of existing mechanisms of space management.

5

MILITARY SPACE
Expanded Uses and New Risks

Military purposes have been part of national pursuits in space since the beginning of space activity. Just as all countries undertake defensive activities on Earth, nations have sought to further their national security through the use of space assets. The fact that relatively few dedicated space weapons have been tested to date and even fewer deployed suggests either that the technology to deploy them efficiently has not yet been developed or that countries have chosen not to do so for political, strategic, or environmental reasons. Analysts are divided over which of these explanations is correct. But the answer matters to the future, because if countries believe that large-scale weaponization of space is inevitable, they are not likely to agree to halt such efforts. By contrast, if the world's major spacefaring countries believe that space weapons are likely to do more harm than good, they are more likely to work to develop restrictive treaties and establish new types of international verification to enforce them.

To date, most military space activities consist of support functions—that is, technologies that allow military forces on the ground, at sea, and in the air to operate more effectively. These include weather forecasting, communications, precision timing and navigation, reconnaissance (of various types), and early warning. Space assets make military systems work better and thereby enhance the tools that can be used in other environments, including improving weapons accuracy to reduce casualties and collateral damage. Reconnaissance and early-warning technologies can help support arms control and prevent conflict in the first place by providing accurate data on long-range delivery systems, thus making it more difficult to cheat. The advantages that the United States gained in many areas of military space during the Cold War have enabled it to project power and preserve peace much more effectively than any other country. Although Russia still possesses significant capabilities, it is further behind the U.S. military in space than it was during the Cold War.

But a variety of other countries are now seeking to enter the military space realm. Most are deploying technologies for reconnaissance, secure communications, targeting, electronic intelligence, and space situational awareness. A few are trying to develop space weapons capabilities, ranging from electronic jammers that interrupt signals to kinetic weapons that destroy spacecraft. Why would they want to do this? The goals of such programs have historically been to try to deny an adversary's "eyes and ears" in space, which could be extremely useful in a conflict. But, as seen in the Soviet anti-satellite (ASAT) test program from 1968 to 1982, the U.S. ASAT test in 1985, and China's ASAT test in 2007, kinetic programs (those based on weapons that collide with their targets) have significant negative implications for the space environment, since they put all other satellites in the same orbital band at risk from the indiscriminate debris they produce. Still, having a potential capability can be considered to serve a deterrent role, making the adversary think twice before engaging in a conflict if there is a risk of losing one's critical space assets. As judged by debates in India since China's ASAT test, the mere fact of having conducted a successful test is perceived as putting one's military space program in a privileged class above others, providing power and prestige.[1] But these very factors make arms control to halt development of such weapons difficult, as countries may be unwilling to sign away options that others have demonstrated, even though none of these weapons have ever been used in warfare and would create long-lasting environmental damage to Earth-orbital space.

In order to understand current trends in the military space realm, it is important to examine what capabilities exist and what pressures might cause an arms race to occur in space, as was threatened at various times during the Cold War (but never took place). Will emerging conditions and the presence of multiple actors make the military space environment more threatening, or will costs, technological limitations, and a desire to pursue peaceful space development trump the use of force and again cause countries to step back from crossing this threshold?[2]

MILITARY MISSIONS AND SPACE PHYSICS

As we saw in chapter 1, the physical characteristics of space greatly influence what kinds of activities are best suited for orbital space and also what they cost. Put simply, the fact that satellites in low-Earth orbit must travel at speeds exceeding 17,000 miles per hour makes them expensive to launch and means that *many* of them are needed in a formation (or "constellation") in order to provide coverage over any specific area of the globe at any given time. Reconnaissance is best done close to Earth in low orbits, but very few countries can afford to orbit enough satellites to make timely passes over single points of interest on the globe more than a few times a day. As in other areas, you get what you pay for, at least if you want to control the information and how often it is provided. Increasingly, commercial imaging satellites can provide "good enough" pictures for those who lack the funds for their own reconnaissance or who want to supplement their limited assets. But national militaries may or may not be able to keep their purchases secret from the governments that have jurisdiction over the satellite companies taking the pictures, and the images may not be as timely as a military purchaser might like.

Beyond photographs, which are now transmitted digitally to computers on Earth and sent to clients electronically (rather than deorbiting film and recovering it manually as in the old days), military customers are often interested in other information: infrared images (which detect the heat signatures of objects) or radar images (which provide information on materials and construction).[3] Unlike visual imaging systems, infrared technologies can work in darkness and radar satellites can operate in rain or shine.

Other information of interest to militaries includes signals intelligence, which captures electronic emissions from radar as well as telemetry from missile tests. Such information is critical for militaries seeking to determine what foreign facilities might target their aircraft with missiles in a conflict and what the capabilities of these systems might be. Signals intelligence satellites may also intercept wireless communications of various sorts.[4]

Medium-Earth orbit is home to most position, timing, and navigation satellites. These satellites are of special interest to military services because of the importance of knowing where enemy (and your own) forces are and where they are going. During wartime, when a military needs to target enemy forces or a specific facility with great precision, these systems are extremely valuable. Such technologies have revolutionized the way the United States fights, giving it stark advantages in accuracy and effectiveness. Despite the fact that the U.S. GPS system is available free of charge to any user around the globe, a number of foreign militaries are seeking to develop their own networks for fear that Washington might shut off or encrypt GPS signals during some future conflict. They may also wish to develop more precise locational signals for their particular region. Russia already has this capability through its GLONASS satellite constellation, China's Beidou system is now reportedly operating on a regional scale, and India, the European Union, and Japan are all in the process of developing such networks.

Geostationary orbit is the location of a variety of other military satellites. Much useful information can be gleaned with various sensors when they are parked above a country (or region) and can "stare" at these locations for long periods of time. While it is not a favorable venue for taking images (due to its great distance from Earth), it is perfect for fixed communications, missile early warning, nuclear test detection, and certain types of signals intelligence.

SPACE WEAPONS

Any consideration of the question of space weapons raises the issue of definition. Some observers make the case that space weapons are already widespread and range from jammers that interrupt the functioning of satellites to past systems like the U.S. space shuttle (which was capable of taking satellites into its cargo bay for repair or, in theory, destruction) to devoted kinetic ASAT systems.[5] Such a broad definition makes space already "weaponized" and renders notions of a "ban" on space weapons impossible. Other analysts argue that only technologies that physically damage or destroy space assets should be counted as "weapons."[6] Far fewer of these latter systems have been developed, and very few of them tested. Prohibiting space weapons through a ban on use, deployment, and future tests may be possible, since destructive testing is highly transparent in space. But no such treaty would be perfect. Non-destructive tampering systems would be harder to limit and would likely require some form of space-based monitoring. Yet the process of elaborating such limits in itself might well be useful, particularly if it improved transparency and provided leverage and incentives for countries to blame, shame, and sanction, thereby raising the costs considerably for violating weapons non-testing or non-deployment norms.

Before discussing national capabilities and specific existing and emerging threats, it is worthwhile to review some basic factors that would affect any country's deployment and use of space weapons. A detailed study of space security by a group of physicists for the American Academy of Arts and Sciences sets forth key parameters that affect military space activities,[7] among them the following:

1. "Satellites are intrinsically vulnerable to attack and interference. However, satellite systems can be designed to be less vulnerable than the individual satellites that compose the system."
2. "A nation could not use space-based weapons to deny other countries access to space, although it could increase the expense of such access."
3. "No country can expect to have a monopoly on deployed ASATs."
4. "Being the first to deploy space-based weapons would not confer a significant or lasting military advantage."[8]

The report goes on to explain some of the technical details about the difficulties of deploying space weapons for various purposes: space to Earth, Earth to space, and space to space. In the first category, the most significant hurdles include the high cost of launching weapons into space and operating them, as well as the need to orbit large numbers of them in order to be able to stop ground-based missiles or attack specific ground targets in a timely manner. Except in distant GEO, weapons do not "sit" above a country but instead fly over very rapidly and then have long gaps between revisit times. This "absentee" problem makes orbital systems more costly and less reliable than air-, sea-, or ground-based systems, particularly for militaries with long-range missiles and GPS access for accuracy.

In the second category, Earth-to-space weapons, the report notes the limited ability of ground-based systems to attack orbital space assets, since specific assets come within range only periodically. Attacking space assets operating in sunlight is relatively easy with an infrared seeker because the target appears warm against the cold background of space. Less-sophisticated optical seekers work too, since the target appears bright against the blackness of space. Attacking in darkness is more difficult, thus limiting the timing of possible attacks, or forcing the use of more easily spoofed radar seekers. Once an attack began, a country whose assets were being attacked could move its space assets to avoid passing over the aggressor country (dependent on timely warning and communications), while potentially taking offensive counter-measures against the attacker's ground systems. Debris accumulation in low-Earth orbit would also quickly become a problem for both sides. Primitive attackers would be unlikely to hit specific assets and would succeed only in spreading debris (or possibly electromagnetic pulse radiation in case of nuclear use) that would be harmful to all countries, thus bringing international scorn and possibly concerted counterattacks.

Lasers—possessed by many countries and used widely for the purposes of determining the altitude of satellites—could be effective in blinding critical assets short of destruction. They can overload the pixels on imaging satellites and cause them to register useless images or even cause permanent damage to their imaging capability. Higher-power lasers could also cause destruction, such as by overheating a satellite and causing a fuel tank to explode. But certain countermeasures can reduce the effectiveness of lasers. For example, some sophisticated military satellites can reportedly detect attempts at laser interference and protect their focal arrays with shutters.[9] Rotating a satellite can also reduce the heating effects of high-powered lasers. In addition, high-power, ground-based laser facilities typically use large quantities of liquid chemicals to power their systems, making them highly vulnerable themselves to cruise missile or aircraft strikes. Overall, Earth-to-space attacks are feasible but are likely to have limited effectiveness, while creating costs for all spacefarers.

In the third category, space-to-space weapons, which are the least developed today, problems arise in terms of useful range, collateral damage, and countermeasures. While it is possible to place a kinetic or explosive space mine in similar orbit as a target satellite, such behavior would likely be transparent in low-Earth orbit and could be difficult to maintain if evasive action were undertaken (depending on which satellite had more fuel). The debris created from an attack would also pose a hazard for all satellites passing through the same orbital band, including those of the attacker. Shooting at targets with a laser would require launching significant amounts of chemical fuel, which would dramatically raise costs for the attacker and possibly explode the target and generate harmful debris. Thus, while a variety of space-to-space weapons options exist, they are technically difficult, relatively transparent (likely resulting in immediate political, economic, or military countermeasures), and costly. Nevertheless, the possibility of space-to-space attack cannot be ruled out, particularly with non-kinetic systems that cause less-than-catastrophic damage (where attribution can be more complicated).

In this regard, another potential means of nullifying a satellite is to jam its signal with electronic systems based in space, in the air, on ships, or on the ground. This method has the great advantage that it does not destroy the satellite, which would likely bring legal or military action by its owner. (To date, no country has destroyed any other country's satellites, only their own in tests.) Interference is normally achieved through electronic means—that is, by sending a signal that overrides the satellite's intended commands, disrupts its receipt of a coherent signal, or prevents its signal from being received on the ground. Such activity represents a violation of commercial satellite regulations, and countries have been reported to the International Telecommunications Union for engaging in jamming. But jammers are widely available in the international marketplace and can be highly effective against commercial systems and satellites that broadcast on fixed, known frequencies. Jamming is less effective against sophisticated military communications satellites, which may incorporate evasive systems that allow them to change frequencies and thus avoid jamming. Compared to ground-, sea-, and air-based systems, orbital jamming is by far the least-developed option. The attacking satellite would have to maneuver into a blocking position relative to Earth and maintain it, as well as have adequate technology to avoid frequency hopping or other evasive measures by the target satellite, again raising costs. But the problem of jammers is likely only to increase as more countries acquire systems for use against civilian and military communications satellites, as well as precision navigation signals.

DEFENSIVE SYSTEMS AND CONCEPTS

While space assets are vulnerable because of their fixed orbits and relative ease of tracking (at least for moderately sophisticated attackers), space systems can also be defended in various ways. A country being threatened, if it has reliable intelligence, might preempt the attack in the first place on the ground. However, that would require an act of war, making it unlikely, unless a conflict had already begun. If it suspected targeting of a specific asset—such as a large intelligence satellite—it might be able to engage in a maneuver to avoid interception. Such a step might be effective if taken early enough, by moving the target spacecraft out of range of a specific country's ground-based missiles, especially if they are being launched from fixed sites. If such a warning were not available, however, a country would find it considerably harder to evade an ASAT attack from the ground. Even short of attack, if the threat of ASAT attack were to cause disruption in an adversary's space constellation by forcing it to take preventive action, it may have achieved at least part of its objective by deterring overflight of sensitive sites.[10]

In sum, while there are some mechanisms to reduce vulnerability, the first shot is still likely to be successful if undertaken by a well-tested ASAT system. Sustaining such a campaign against multiple spacecraft, however, is much more difficult. Fortunately, not many ASAT weapons have undergone multiple tests or are readily deployed in significant numbers.

U.S. anti-ballistic missile defenses—including those based on Aegis ships—have had the most operational testing of any system with either a devoted or a dual-use ASAT capability. Foreign critics of U.S. space policy raise this point frequently, and it does complicate space arms control efforts. While there are differences between satellite and ballistic missile interceptions, missile defense is generally harder. Missiles come in various sizes and speeds, and their warheads that travel through space are typically much smaller than satellites. Also, satellite orbits can be observed beforehand, and the spacecraft themselves are often large and reflect sunlight well, making them easier to target. Finally, satellites offer multiple passes, allowing an attacker to prepare for the shot over days and weeks. In missile defense, there is only one shot, whether the intercepting country is ready or not.

In seeking to defend one's satellites, employing so-called non-offensive defenses may be the most effective strategy for both deterring attacks against satellites and preventing them once a conflict begins. As the Canadian space expert Phillip J. Baines explains, these options include five categories: (1) denial and deception (for example, the use of black, carbon-impregnated thermal blankets to mask a satellite's optical signature); (2) hardening and shielding (for example, onboard shutters to protect against lasers); (3) maneuvering (for example, the addition of stored fuel for possible evasive actions); (4) redundancy and reconstitution (for example, the use of commercial or allied systems to provide service in case of attack on one's own assets); and (5) dispersion (for example, the creation of small, hard-to-attack modular satellite networks to replace constellations of large and vulnerable multipurpose satellites).[11]

At present, no country is known to have any significant stockpile of ready ASAT weapons. Thus attackers face a significant deterrent if they consider that their limited weapons will be rendered irrelevant by defensive countermoves. In addition, retaliatory strikes on the attacker's ground systems (launch sites, communications nodes, radars, and command/control)—once a conflict started—could be swift and devastating, given the fragility and vulnerability of many of these installations.

Short of direct attacks on space systems, space war could occur through several other means. Indeed, as space powers know, the most vulnerable part of most constellations is not the space-based portion. Existing ASAT systems, as noted above, are relatively expensive, few in number, and of limited effectiveness and reach. By contrast, ground-based radars, launch sites, control facilities, and communications nodes, as well as the radio signals themselves, are often vulnerable to simple conventional attack. However, since these systems are located in another country's national territory, the threshold of war would have to be crossed for such an attack to be undertaken.

Possible Space War Scenarios

What might a space war look like? It could begin with an increase in tensions over terrestrial issues and build into a conflict in which space assets (such as those for navigation, reconnaissance, and targeting) would play a critical role and might quickly become desirable targets. An advanced space power might target an enemy's satellites in low-Earth orbit and seek to destroy them with kinetic weapons in order to "blind" enemy forces. Such attacks would release large amounts of debris, quickly putting other countries' spacecraft in the same orbits at risk. A less-developed country with little reliance on space might seek to carry out an even less discriminating attack using a nuclear weapon, although at the risk of destroying or disabling all satellites passing through that region of space. The country that had been targeted in space might respond by trying to take out launch sites in the aggressor country, causing casualties on the ground as well as destruction of critical facilities. This would be more difficult with mobile launchers, although the strikes could focus instead on vulnerable command and control sites. Space assets of that country might also be targeted for counterattack, increasing the field of orbital debris even more. Once these attacks are carried out, that country's launch sites might then come under attack, escalating the conflict even further. At this point, the war might spiral out of control as each side sought to take out more facilities and troops to render space unusable to its opponent. Depending on the armaments involved, this could lead to limited nuclear exchanges. Alternatively, the rapid rise in lethality and damage might bring the two sides to their senses and cause the two capitals to seek a cease-fire. But wars often make people mad, rendering sensible outcomes less likely.

A less escalatory but effective means of negating an adversary's space systems is ground-, air-, or possible space-based jamming of the adversary's satellite signals. Of these different technologies, ground-based systems are generally cheap and fairly effective, at least for temporary effects. The dilemma, though, facing countries that might attempt to jam GPS signals and communications in a theater of conflict is that the more advanced militaries already train to operate under degraded conditions. The United States, for example, is developing systems to reroute GPS signals and distribute other communications through nontraditional means. Thus, a major investment in anti-space technology could be rendered moot. That said, it is undeniable that valuable space assets could suffer significantly from a concerted attack by a peer or near-peer adversary, reducing overall combat effectiveness. However, deterrence might also prevail. The United States considers attacks on its critical military space systems as possible precursors to nuclear war. Thus, the attacker would have to think very carefully about the risks of undertaking such actions.

Indeed, one of the stabilizing characteristics of the Cold War was the understanding that attacks on strategic space reconnaissance and early-warning systems could indeed be interpreted as the first stage of a nuclear attack. For this reason and because of the need for such systems to verify U.S.-Soviet arms control agreements, the two sides agreed by treaty to exempt such systems from attack and by consensual norm not to interfere with each other's military satellites more generally. A question for the future, however, is whether this understanding has been adequately instilled in the minds of officials among new space powers. In this regard, political and diplomatic mechanisms provide another important line of defense in space.

NATIONAL MILITARY SPACE CAPABILITIES

Dedicated space weapons have been developed only by Russia, the United States, and China to date, with relatively limited capabilities. The one exception is nuclear weapons, which represent a blunt and powerful anti-space instrument that a number of other nuclear- and missile-capable countries might use against space if they wanted to cause indiscriminate damage to all space assets. Nuclear weapons have been used only in U.S. and Soviet test programs early in the space age. Since then, no such tests have occurred in space and almost all space-faring countries have agreed to the terms of the 1963 Partial Test Ban Treaty or the 1996 Comprehensive Test Ban Treaty and have sworn off the use of nuclear systems. Of course, it remains possible that a country might violate these agreements. But the threat of destruction of large numbers of satellites and the deaths of astronauts in low-Earth orbit should serve as a significant disincentive, unless a country is willing to bear the wrath of all countries with space assets. Conflict in orbit, however, remains difficult to predict. In part, this confusion has to do with the broad range of military capabilities present among current and emerging space actors.

Today's dynamics pose a different kind of military space competition than the one that existed during the Cold War. It is slower and more diffuse, but it is beginning to accelerate. The big question is whether national militaries will by and large limit themselves to military support activities and force enhancement technologies or will instead venture into costly and provocative force-application programs for space.

The United States

The United States' military space program is the most comprehensive in the world, dwarfing all others, including those of Russia and China. According to published sources, combined U.S. spending on military and intelligence activities in space is about $42 billion a year,[12] a figure that surpasses the combined figure for all other world military space programs. Of the approximately 1,000 operating spacecraft currently in orbit, some 170 are military satellites and about half of those are operated by the U.S. military and intelligence communities.[13]

The United States has a long history of experimenting with offensive and defensive space systems. Notably, very few of these technologies have been deployed in anything beyond "hedge capability" numbers because of cost, concerns about strategic stability, and calculations of their likely limited operational effectiveness. Nuclear-tipped ballistic missile defense (BMD) systems for use in space, nuclear-tipped ASAT weapons, and air-launched kinetic ASAT weapons have all been tried and abandoned. Offensive U.S. capabilities for use against foreign space assets are limited largely to dual-use capabilities from programs with other primary uses. The sea-based Aegis BMD system proved effective in destroying a falling and unresponsive U.S. satellite laden with hydrazine in February 2008 and presumably could be used again. U.S. laser-ranging facilities in New Mexico have dual-use capabilities as dazzlers (capable of temporarily blinding foreign reconnaissance satellites) and could perhaps have destructive (permanently blinding) capabilities, depending on their power and the length of time a target could be engaged.[14] The U.S. military has also experimented with satellite proximity operations: the 2005 Demonstration for Autonomous

Rendezvous Technology (DART) project, during which the test vehicle inadvertently collided with the target satellite, and the *Experimental Satellite System (XSS)-11*, first tested in 2005–2006.[15] Analysts also presume that the U.S. military has satellite-jamming capability "effective out to geo-synchronous orbit," although the United States is not known to have operated such capabilities.[16] The U.S. military also has sophisticated satellites equipped for such missions as signals intelligence, synthetic aperture radar detection, optical detection, and infrared missile detection. It is the only country to date to have used GPS-guided weapons in war. The U.S. military has tested a small, experimental, unmanned space plane (the *X-37B*), apparently for the possible delivery of weapons or small satellites, space-based equipment testing, or military reconnaissance.[17] U.S. missile defense programs, ranging from ground-based interceptors in Alaska and California to mobile sea-based Aegis destroyers, also have an inherent dual-use capability as possible ASAT weapons.[18]

Russia

Russia is the next most capable military space actor, based on its long history of military space operations during the Cold War. While many of these capabilities deteriorated in the 1990s, the Russian government has sought to reconstitute a number of them. These include the direct-ascent Naryad ASAT system (although not tested in a destructive mode), the GLONASS precision navigation and timing constellation, and various signals intelligence, photo-reconnaissance, communications, and meteorological satellites. Moscow tested co-orbital ASATs during the Cold War and has conducted experiments with other space weapons. Russia today has laser facilities and is known to produce satellite jammers, some of which were sold to Iraq before the second Gulf War in 2003 in an attempt to negate the U.S. GPS system (but they were destroyed by U.S. forces).[19] In his third term in office, President Putin has pledged major increases in military space spending. Moscow recently modernized the Plesetsk military launch site to increase Russia's capacity to reconstitute its former military satellite constellations.

Also, various organizational reforms have placed more emphasis on Russia's Space Forces, which number some 40,000 members.[20] Still, the major questions facing Russian military space capabilities are ones related to research and development, quality control, and long-term political and budgetary support.

China

China is a relative newcomer to the military space club, but it has been making up for lost time. In the past decade, it has made major investments and conducted tests of counterspace systems, including kinetic ASATs, small satellites capable of proximity operations, and jammers. China is also known to have laser facilities capable of disrupting space assets. A survey of Chinese military space writings by Dean Cheng argues that "PLA authors . . . would seem to support an approach that balances disruption (soft-kill) and destruction (hard-kill) of an opponent's space systems."[21] A 2011 U.S. government report makes the case that, besides Beijing's 2007 ASAT test, "China is also developing other kinetic and directed-energy (e.g., lasers, high-powered microwave, and particle beam weapons) technologies for ASAT missions. Foreign and indigenous systems give China the capability to jam common satellite communications bands and GPS receivers."[22]

Besides these programs, analysis of recent launches suggests that the bulk of China's military space expenditures has gone to expansion of traditional military support capabilities, as Beijing seeks to catch up with the United States and Russia and to develop assets that will be useful for modern combat operations and global force support. China has focused on expanding the size of its constellation of reconnaissance satellites, as well as improving their previously poor resolution, while also developing new radar satellites and expanded space-based electronic intelligence-gathering (another earlier weak point).[23] Nevertheless, China continues to purchase commercially available visual spectrum and infrared imagery, suggesting that gaps remain or that the resolution of Chinese military satellite technology is not yet adequate.[24]

China is also moving rapidly to populate its Beidou precision timing and navigation satellite network, which will likely have a separate military signal for use in missile guidance. The question that observers watching China's military space expansion ask is, "Where exactly is this program headed?" Some experts believe that China is seeking a limited "hedge" capability to enable it to deny possible U.S. space dominance in case of a conflict over critical national interests, such as the status of Taiwan, which China claims as an integral part of its territory. As the military space expert Barry Watts argued in 2011 testimony before a congressional commission, U.S. fears of a "Space Pearl Harbor" proved a poor predictor of China's military space aims during the decade from 2000 to 2010.[25] While not ruling out future expansion of Beijing's capabilities, Watts concluded that, overall, China's military space efforts "would be unlikely to produce a decisive advantage over the United States in conflicts in the western Pacific through the end of this decade," and even less so at the global level.[26] But others believe that China's military space growth is aimed at developing options for full-scale space war. One analyst of Asian affairs, Gordon Chang, argued in 2009 that Beijing had "announced its intention to begin the space arms race in earnest" and had adopted a policy to "dominate space."[27] Thus far, the evidence seems to point to more-limited Chinese aims in the space weapons sector focused on developing deterrent capabilities and limited offensive systems, rather than a full-scale war-fighting arsenal for space. But time will tell.

The next tier of space actors—as a group—devotes far fewer resources to military space activities than do the top three. Thus far, they have tended to limit themselves almost exclusively to support operations, such as reconnaissance and communications. But this situation is beginning to change.

European Space Agency Countries

Several countries in the European Space Agency (ESA) have military space activities, with France being the most experienced. These activities have historically been conducted strictly on a national basis because of

ESA's original charter requirement that joint activities have solely civilian purposes. France has used its Satellite pour Observation de la Terre (SPOT) system for military reconnaissance as well as civilian remote sensing. It followed with two generations of higher-resolution *Helios* satellites, with Germany as a partner. Since 2012, an even more sophisticated Pleiades satellite constellation has provided 70-centimeter-resolution images to France's military, but also sells imagery commercially.[28] In addition, France continues to develop its space-based missile early-warning system. Germany operates highly capable synthetic aperture radar satellites under the SAR-Lupe program, which has a ground station in France as well. Italy operates the Constellation of Small Satellites for the Mediterranean Basin Observation (Cosmos-SkyMed) radar system, whose data it swaps with France for optical imagery. These countries are being joined by Belgium, Greece, Italy, and Spain in working toward the Multinational Space-Based Imaging System (MUSIS).[29] Given the high costs of national systems, MUSIS is an effort to pool resources and share data from various national platforms and ground stations. Finally, the United Kingdom has long operated a military communications system (Skynet) and is working toward greater cooperation in the military space sector. Recently, as a result of pressure from member governments and in the context of Europe's planned Galileo GPS system, ESA nations agreed to allow joint military activities. European militaries plan to equip various defensive and offensive systems with Galileo devices to provide precise tracking and targeting.

Discussions are also ongoing within the context of the NATO alliance to begin operational cooperation in some areas of military space, thus reducing the barriers that have long existed, even during the Cold War. Part of the reason is cost, but the increasing use of space assets in military operations requires greater cooperation if alliance effectiveness is to be maintained and expanded. U.S. military officers have largely abandoned the go-it-alone mentality of the Cold War period and recognize the advantages of positioning the United States in a leadership role among other like-minded countries in space. As General James Cartwright (U.S. Marines), at the time the vice chairman of the Joint

Chiefs of Staff, argued about military space cooperation at a national conference in 2011, "We can't afford these constellations ourselves."[30] He also noted the reality of "coalition warfare" in the modern age and emphasized the importance of breaking down secrecy barriers among allies in space. Otherwise, he said, "it's like having a guy in the foxhole with you who's not armed."

India

The expanded use of space by the world's leading militaries has not gone unnoticed by their rivals. Following China's 2007 ASAT test, India announced the formation of the Integrated Space Cell to coordinate a series of new efforts to make greater use of space assets for military purposes.[31] Indian officials also stated that they would match China's ASAT capability either through a kinetic missile defense interceptor or through ground-based lasers. This announcement shows the salience of "tit for tat" arming in regard to regional space dynamics. In recent years, India has teamed with Israel to acquire highly accurate satellite reconnaissance technology, although civilians have operated these services.[32] Given the Chinese challenge, however, India has now directed its civilian space agency to build dedicated military satellites for each branch of the Indian armed forces. For the first time in India's history, military personnel will operate these satellites. While figures for India's military space budget are not published, it is likely that defense efforts will at least match the double-digit increases in India's recent civil space budget as New Delhi struggles to remain competitive with China.

Japan

Japan is another recent and quite unlikely entrant into the military space realm. In the late 1990s, Japan reacted to North Korea's Taepodong-1 missile test (which overflew Japan before its third stage failed) by authorizing the country's first photo-reconnaissance system: the Information-Gathering Satellites (IGS). However, because a 1969 space law limited the country's space activities to civilian purposes, it was necessary to create a separate agency under the Cabinet Secretariat to manage this program. To the surprise of many outside observers, the Japanese legislature moved further toward military uses of space in reaction to China's ASAT test by passing a long-proposed reform of the 1969 law. This 2008 legislation allowed the use of space for military purposes. Japanese officials indicated that even space weapons might be allowed, as long as such systems were "defensive" in nature.[33] The military's possession of both Patriot and Aegis BMD systems creates at least a potential ASAT capability for Japan, although the military has never tested such systems against space objects or, as far as observers know, configured the system's complicated software for that purpose. But Japan's space industry is pushing strongly in the direction of expanded military space activity (including possible space-based defenses), seeking lucrative contracts to expand its long-stagnated domestic market.[34] Japan is also investigating possible development of a satellite-based early-warning system to detect foreign missile launches, despite the high costs. The country's biggest problem in seeking to maintain its place as Asia's technological space leader is budgetary.

Additional Countries

Among other countries with military space programs, relatively few have the capability of building their own space assets. Israel is one exception, having long operated *Ofeq* high-resolution reconnaissance satellites, whose services and technology it has shared with such partners as India and Taiwan. Putting a premium on military-technological independence, Israel has also developed synthetic aperture radar satellites. North Korea, by contrast, has thus far proven incapable of developing modern satellite technology domestically, and its pariah status has hampered aquisition aims. But another group, which includes Australia, Brazil, Iran, Singapore, South Korea, Vietnam, and a range of others, are using a combination of foreign and some domestic sensors to develop military support programs. One advantage for today's late-developing space actors is that they can purchase foreign commercial imagery and bandwidth on commercial communications satellites while accessing freely available U.S. or other global positioning signals to assist in their military operations. The only risk is that access to some of these technologies might be cut off in times of war.

DEBATES OVER MILITARY SPACE STRATEGY AND POLICY

Despite the spread of military space capabilities, the sky is not falling and destructive space conflicts have not emerged. Self-interest has acted as a powerful constraint, at least in terms of deliberate tampering with, damaging, or destroying foreign space assets. Jamming has, however, occurred with increasing frequency and is almost inevitable in the context of possible future warfare. Some believe that kinetic weapons are inevitable too, although no country (except arguably the United States in the context of its missile defense program) has any significant number of weapons ready for possible use against space-based systems (and some modifications would be required). China could certainly expand the number of its mobile missiles equipped with ASAT seekers. Russia could do the same. Other countries might follow suit, if the leaders were to move in this direction.

Military strategy and policy are the final part of this equation that requires further analysis. How do countries see the future of military space activity and what factors are likely to guide their relations? Is conflict prevention, or at least management, possible? Few countries publish official space policies that cover their civil *and* military aims and intentions. The United States is the one major exception and has called on other countries to do the same.

In 2010 the United States issued a National Space Policy (NSP) that both reaffirmed past approaches (such as the inherent right to self-defense in space) and broke new ground in terms of its outlook toward the international space community. In part, the 2010 NSP represented a reaction to the 2006 NSP issued by the George W. Bush administration, whose go-it-alone approach to military space and rejection of new arms control or other legal mechanisms alienated other nations. The underlying assumption of the 2006 NSP was that in the face of expected future foreign threats, the United States needed to investigate a range of possible space weapons and deploy the ones it believed most effective to prevent or prevail in an inevitable space conflict. Such assumptions came out of the 2001 Rumsfeld Commission report on the management of U.S. space assets, which warned of dangerous U.S. vulnerabilities. Moreover, as Undersecretary of the Air Force Peter B. Teets argued in the introduction to the Air Force's *Counterspace Operations* guidelines issued in 2004: "Controlling the high ground of space is not limited simply to protection of our own capabilities. It will also require us to think about denying the high ground to our adversaries. We are paving the road of 21st century warfare now. And others will soon follow."[35] Building on these assumptions, the 2006 NSP identified a logical evolution among concepts of sea, air, and space "power" over time. But the experience of China's 2007 ASAT test, the absence of a U.S. political strategy for space, and further considerations of the harmful global implications of any space warfare led the Barack Obama administration to try a different tack.

The Obama team did not focus on inevitable space conflict and insist on complete U.S. freedom of action, which it saw as stimulating the development of foreign space weapons, signaling a tolerance for their space-weapons tests, and accepting the pollution of low-Earth orbit from increasing orbital debris. Instead, President Obama's policy advisors took a page from the Kennedy administration and decided to step back from the precipice of a seemingly brewing arms race. They outlined a "collective responsibility" approach to space security in an effort to halt what they saw as a dangerous and preventable trend by refocusing international attention on *shared* interests in safe access to space. As the 2010 NSP explained, "The . . . interconnected nature of space capabilities and the world's growing dependence on them mean that irresponsible acts in space can have damaging consequences for all of us."[36] With this in mind, the administration called upon all countries "to work together to adopt approaches for responsible activity in space to preserve [safe access to space] for the benefit of future generations."[37] While the 2010 NSP did not call for specific new treaties, it did renew the U.S. commitment to international cooperation in the pursuit of enhanced space stability through innovative partnerships, including working with "civil, commercial, and foreign partners to identify, locate, and attribute sources of radio frequency interference."[38] It also identified enhanced military space cooperation with allies and "bilateral and multilateral transparency and confidence-building measures" with others "to encourage responsible actions in, and the peaceful use of, space."[39] Internationally, the 2010 NSP received little of the criticism that greeted the 2006 U.S. document, and much praise.

Building on this foundation, the Obama administration issued a first-of-its-kind National Security Space Strategy (NSSS) in January 2011 that outlined how the U.S. military and intelligence communities would implement the new NSP. Instead of emphasizing the use of force in space and calling for deployment of U.S. space weapons, it sought to raise the bar, describing a vision of a cooperative environment that would benefit all users. This vision is worth quoting at length, as it represents a unique effort by the U.S. military to state clear objectives for all countries in space focused on restraint, communication, and cooperation:

> We seek a safe space environment in which all can operate with minimal risk of accidents, breakups, and purposeful interference. We seek a stable space environment in which nations exercise shared responsibility to act as stewards of the space domain and follow norms of behavior. We seek a secure space environment in which responsible nations have access to space and the benefits of space operations without need to exercise their inherent right of self-defense.[40]

In terms of working with others, the 2011 NSSS took a forward-leaning approach to international outreach, seeking to change traditional norms in space security affairs of non-communication, secrecy, and a focus on national-technical solutions. Part of the reason, clearly, was the failure of past policies to prevent events like the Chinese ASAT test and the collision of a U.S. *Iridium* and a Russian *Cosmos* satellite in 2009. The aim of the new approach was preventive, not simply reactive.

The NSSS section on deterrence of aggression against so-called space infrastructure stated that Washington would

> support diplomatic efforts to promote norms of responsible behavior in space: pursue international partnerships that encourage potential adversary restraint; improve our ability to attribute attacks; strengthen the resilience of our architectures to deny the benefits of an attack; and retain the right to respond, should deterrence fail.[41]

Overall, the 2011 NSSS emphasized a "multilayered deterrence approach" that put military means as the last resort and sought to exercise a range of economic, political, and diplomatic options to prevent conflict. Instead of terms like "space control" and "space dominance," the new U.S. approach stated in its conclusion: "Our objectives are to improve safety, stability, and security in space," and to work toward "creating a sustainable and peaceful space environment to benefit the world for years to come."[42] The document helps shift international attention toward diplomatic solutions, although much work remains to be done in order to bring the lofty visions of the 2010 NSP and the 2011 NSSS to fruition.

Critics within the United States, however, believe that the administration has begun to surrender the U.S. advantage in military space by failing to continue some of the more aggressive programs (such as the space-based laser, kinetic-kill interceptor, and Brilliant Pebbles) all revived in the Bush administration (although later denied funding by Congress for both technical and financial reasons). One critic, Everett Dolman of the Air Force's School of Advanced Air and Space Studies, writes of the inevitability of a coming war with China in space and the need for the United States to abandon collective security for a space dominance strategy based on control of low-Earth orbit through the deployment of orbital weapons.[43]

Notably, the current U.S. Defense Department counterpoint paints a very different picture of the future based on the potential ability of leading spacefaring nations to prevent conflict. An essay in 2012 by the head of Space Policy in the Office of the Secretary of Defense and one of his advisors puts future space security into an environmental context, emphasizing the need to "address the challenges of a domain that is increasingly congested, contested, and competitive."[44] They call upon all nations to develop a "common space 'rule set'" to allow "military space operators and intelligence analysts to more easily identify irresponsible actions by aggressive or rogue actors, enabling accurate attribution and possibly building consensus for coalition or international action to uphold freedom of access to the space global commons."[45] Their concept focuses new attention on the commercial notion of "best practices" for space and takes a preventive approach to possible conflicts. Rather than placing a priority on developing and immediately deploying large constellations of space weapons, they state that "broadly increasing dialogue between space-faring nations can help build understanding and strengthen relationships that could prove invaluable during a potential crisis."[46]

The question going forward is whether countries pursuing military space programs and possible weapons in the context of regional competitions will prove receptive to global notions of "responsible behavior" and "best practices." History suggests that countries will act selfishly and will cheat on agreements if given the chance. But the U.S. ability to attribute to specific actors dangerous space behavior through its Joint Space Operations Center and increasingly accurate network of radars and other sensors—possibly with international input in the coming years—could act as a serious deterrent to potential violators of such norms, at least in major spacefaring countries, such as China, India, and Russia. This "community policing" approach is one that has never before been attempted in space, but it may succeed because of shared military interests in maintaining safe access to the valuable information that travels through space and the unique observations possible from space-based assets.

The satellite non-interference principle has remained part of all U.S.-Soviet and U.S.-Russian arms control treaties since 1972. But it has not been extended to military space relations involving China, India, Israel, Japan, the European countries, or other emerging space actors, thus risking instability as new military space forces develop. If such a norm could be established among the major military space countries, it could promote a reduction in tensions and increase prospects for cooperation. But more serious efforts to stem the development and testing of new space weapons and to foster military-to-military cooperation are likely to be prerequisites for heading off currently dangerous trends, particularly as space becomes an added dimension of festering regional rivalries in East Asia, South Asia, and the Middle East.

CONCLUSION

Several areas in the military space realm merit additional attention in terms of emerging concerns. These include proximity operations, new kinetic systems, airborne or other mobile lasers, and the proliferation of hard-to-monitor micro-satellites. In some cases, there are military countermeasures that could effectively mitigate these threats. In others, the "fixes" might require additional space situational awareness or new political means, possibly including new forms of collective security or space "policing." In some areas of military security, use of the independent scientific and Internet communities could play an important supporting role in verifying compliance and sharing data on wrongdoers.

Overall, space security developments in the twenty-first century provide reasons for both worry and optimism. Military technology relevant to space is spreading to new actors, who may (at least initially) have less interest in preserving space and accepting norms of non-interference with other actors' spacecraft. The use of kinetic space weapons by both China and the United States (albeit against their own satellites) can be viewed either as a harbinger of future conflict or as a warning sign of what we need to prevent.

6

SPACE DIPLOMACY

China's 2007 anti-satellite (ASAT) test exploited a gray area in international space law. The Outer Space Treaty calls for prior notification of other countries in the case of activities that might cause "harmful interference" with the space programs of other countries.[1] Despite the ASAT test's release of thousands of pieces of dangerous debris, no such consultations took place. Chinese officials likely assumed that since the Soviet Union had conducted some two dozen such tests from 1968 to 1982 and the United States had carried out one in 1985—also with no consultations—that China's test would be able to slide under the literal and figurative radar screens. Oddly, the United States, which had observed two prior Chinese ASAT tests that had, intentionally or not, missed orbiting satellites, decided not to request a consultation with China in advance of the third test. Instead, it was amateur astronomers who noticed the disappearance of an aging Chinese weather satellite and brought international attention to this anomalous event, causing the U.S. military to confirm China's destructive action about which both Washington and Beijing had earlier remained silent. With space increasingly crowded by 2007, international condemnation then came loudly from many corners: other governments, scientists around the world, and even private satellite companies.

A year later, in February 2008, the United States decided to destroy its unresponsive *U.S. 193* intelligence satellite, stating that the action was being taken to prevent it from reentering the atmosphere fully loaded with toxic hydrazine (although critics suggested the shootdown was actually a signal to China).[2] Whatever the true reason, Washington conducted the world's first advance consultation under Article IX of the Outer Space Treaty, sending a senior NASA debris expert (Nicholas Johnson) to Vienna to explain the planned U.S. activity and its consequences before member states of the UN Committee on the Peaceful Uses of Outer Space (COPUOS). Thus, despite its prior practice in 1985, the United States decided that evolving conditions in space now made enhanced international transparency the appropriate behavior in order to avoid condemnation. Given the satellite's very low altitude, the debris from the U.S. destructive action deorbited from space within months and posed little danger.

The recent spread of space capabilities to many more nations brings with it a new imperative for making sure all actors behave responsibly. It took the United States and the Soviet Union several decades to work out formal and informal rules for managing the risks of space conflict. This process will now have to take place faster if problems are to be avoided. But it may be difficult for emerging spacefaring nations to grasp or accept policies of transparency and restraint, especially when they are preoccupied with political and military rivalries and bent on achieving advantages over their adversaries.

To date, space is the only environment of human activity (except the Antarctic) that has not witnessed direct international conflict. In legal terms, space has been a realm of shared ownership since the passage of the 1967 Outer Space Treaty. Some scholars have likened space to a "commons," an English institution represented by the shared land historically attached to villages where animals could graze freely.[3] Over time, however, as more and more people brought increasing numbers of animals into these spaces and fodder became scarce, many of these areas became unsustainable. The village commons had to be broken up, usually ending up in private hands. Questions about whether space will be carved up as a result of emerging conflicts over finite orbital and celestial resources continue to concern academics,[4] as well as government officials.[5]

One factor is the increasing military use of space. As discussed in chapter 5, until recently, only the United States and Russia had serious military space programs. In the past decade China has joined them, and a growing number of other countries have begun such efforts. This changing situation raises the possibility of conflict, whether planned or inadvertent, as national military space objectives collide. Civil and commercial crowding of space also means more spacecraft to track, more orbital debris, and more political problems. Governance mechanisms will have to evolve to manage the growth of actors and the spread of space technology.

Space diplomacy has moved in fits and starts to address commonly identified problems. But the treaties and other accords reached by countries in the 1960s and 1970s did not ban military activity altogether. They also left intentional and unintentional loopholes for certain types of weapons, which countries either decided they could not verify via a negotiated ban or wished to leave open for their own possible development. As the leaders, the United States and the Soviet Union kept close track of each other's military test programs and tended to behave cautiously in the knowledge that their adversary would likely respond actively to any attempt to obtain a unilateral weapons advantage. By the 1990s, the United States and Russia shared a strong norm against testing kinetic weapons against satellites, but they did not sign a new treaty attempting to cement this practice into an international legal rule or to extend it to other actors. China eventually exploited this ambiguity. Today, significant gaps remain in international treaties regarding space activity. There are no treaties that prohibit the testing of non-nuclear weapons in space (including kinetic munitions, lasers, electronic jammers, and microwave systems), no restrictions against orbiting such non-nuclear weapons, and nothing but voluntary guidelines to prevent countries from releasing harmful orbital debris. Prior notification is the only emerging norm.

China's rise as a significant military space power challenges the old bilateral (U.S.-Russian) leadership of space security affairs. In Asia, strong nationalism has characterized an evolving regional competition among China, India, Japan, and the two Koreas, among others. Many of these countries harbor deep-seated historical animosities and have no tradition of arms control or security cooperation. In the Middle East, Iran has joined Israel as a spacefaring nation, but Tehran has violated international commercial norms by jamming the signals of certain foreign satellites broadcasting over its territory. In South Asia, Pakistan aims to counter India's recent venture into military space activity through cooperation with China. The weakness of enforcement mechanisms in current international space law and the holes in the space security framework raise serious questions about the adequacy of existing governance tools.

But military tensions alone do not account for the full extent of today's space governance problem. The UN Conference on Disarmament (CD) in Geneva, which is responsible for negotiating international arms control treaties, has been stalemated since the late 1990s by conflicting national priorities and a crippling consensus rule that prevents formal discussions unless all countries agree to go forward. No space negotiations have been held at the CD since the mid-1990s. In fact, no new international arms control mechanisms for space have emerged since 1975.

In order to explain how we got into this situation and how we might move beyond the current impasse, it is worthwhile to review the major directions of late-twentieth-century space diplomacy, identify emerging twenty-first-century trends, and discuss the challenges countries face today in trying to manage space collectively and avoid conflict. These dilemmas are compounded by vast differences in capabilities among the actors, the relatively large role of military activities (some secret) among the top three spacefaring nations (China, Russia, and the United States), and enduring patterns of mistrust in international relations more generally, which make it difficult to reach binding agreements and to enforce them. Nevertheless, unilateral military approaches to space security can go only so far. Relying mainly on weapons to provide security is costly, risky, and escalatory, as these systems often stimulate rivals to develop systems to counteract them, leading to potential arms competitions and the heightening of tensions. These points highlight the important role of diplomacy in any successful space future. The trouble is that to move in this direction, countries have to identify areas of common interest, craft agreements, and rally the political leadership needed to implement the agreements.

BACKGROUND TO TODAY'S DEADLOCK

From 1963 to 1975, the United States and the Soviet Union led efforts to create a basic framework for space security. As described in chapter 2, these agreements included the 1963 Partial Test Ban Treaty, the 1967 Outer Space Treaty, the 1972 Liability Convention, the 1972 Anti-Ballistic Missile Treaty, and the 1975 Registration Convention. But with the breakdown of U.S.-Soviet détente in the late 1970s, the political environment for new agreements evaporated. A series of talks aimed at halting further development and testing of ASAT weapons nearly reached fruition in 1979, but complications in the bilateral relationship introduced by the Iranian revolution and the Soviet invasion of Afghanistan caused this tentative accord to be pulled from consideration.

With bilateral nuclear and space tensions rising, a group of Western countries (led by Italy) that were concerned about the possible extension of the arms race into space joined in an unusual coalition with the Soviet Union in support of a UN resolution in 1981 on the Prevention of an Arms Race in Outer Space (PAROS).[6] With continued Soviet ASAT tests and the ramp-up of the U.S. Strategic Defense Initiative (SDI), which planned to deploy thousands of space-based interceptors for missile defense purposes, these concerns only increased. Given its SDI plans, the Reagan administration resisted international efforts to negotiate a new space treaty. The Soviet Union countered with a surprising proposal in 1983 to halt all ASAT testing and agreed to dismantle its existing ASAT system. But the Reagan administration doubted it would ever be able to determine whether Moscow was complying with its claims, if the agreement were to go forward. After a series of deaths of elderly Soviet leaders, a new Soviet leadership emerged in 1985 under the relatively young reformist Mikhail Gorbachev, who made an even more radical suggestion: the formation of an International Space Authority to ensure the peaceful uses of space and to help verify a new treaty against the weaponization of space. In the end, the United Nations turned these proposed space negotiations over to the CD, where the Ad Hoc Committee on PAROS held discussions (albeit inconclusive ones) from 1985 to 1994 on mechanisms to strengthen space security. The Reagan administration participated with skepticism in these talks and in bilateral space and defense discussions with Moscow linked to the nuclear arms control process. The U.S. government viewed these international space negotiations as an effort to block the SDI program and other missile defense efforts, a policy that largely continued under President George H. W. Bush and President Bill Clinton. After the Soviet breakup in 1991, most countries no longer viewed space conflict as imminent. As a result, the negotiating mandate on space security at the CD finally expired in 1995.

But by the late 1990s, with the U.S. test program for missile defenses beginning to move toward interceptors whose operational altitudes and speeds might put Russia's and China's nuclear deterrents at risk, Moscow joined with Beijing in an effort to renew the CD's negotiating mandate on space arms control. Moscow also sought to limit U.S. missile defenses through insistence on maintenance of the ABM Treaty, although allowing some systems via so-called demarcation agreements, so long as their speed and range did not allow creation of a nationwide missile defense. Such limits met with strong opposition within the Republican-controlled U.S. Senate, many of whose members wanted a U.S. withdrawal from the ABM Treaty altogether to allow progress toward a national (vs. a site-defense) system. The Clinton administration decided to side with Republicans in blocking space talks at the

CD, insisting instead on negotiations for a fissile material cut-off treaty (or FMCT), intended to halt the global production of fissile material for weapons purposes. Absent a consensus on the agenda, the CD talks could not resume, and no talks were held throughout the subsequent George W. Bush administration as well.

Only in June 2009, with mutual Chinese and U.S. compromises to address space security, the FMCT, and other issues, did CD delegations finally agree to a mandate for talks. A few months later, however, Pakistan blocked this new consensus by objecting to the fissile material cut-off talks, thus throwing the CD back into deadlock.

Since the mid-1970s, significant international progress toward enhanced space governance has taken place in only one area: orbital debris control. After the 1985 U.S. ASAT test, the Department of Defense and NASA became increasingly concerned about the threat of orbital debris and began bilateral discussions with allies and eventually the Soviet Union on the need for debris control. These efforts resulted in the formation by the United States, Japan, ESA, and the Russian Federation of the Inter-Agency Space Debris Coordination Committee (IADC) in 1993. This body later expanded considerably and began drafting a set of guidelines for best practices in debris mitigation,[7] including cessation of the use of hazardous devices such as the exploding bolts that used to be released and put into orbit when rocket stages separated. The IADC called upon states to refrain from the creation of long-lasting debris (longer than twenty-five years) and to deorbit low-Earth orbit (LEO) satellites at the end of their service lives and boost geostationary orbit (GEO) satellites to higher, supersynchronous orbits to prevent collisions. The IADC eventually worked with COPUOS to craft a voluntary set of best practices for consideration at the United Nations, which approved the Space Debris Mitigation Guidelines in December 2007 by unanimous consent.

THE NEED FOR BETTER SPACE GOVERNANCE

Given the slow pace of progress toward expanded space governance since the mid-1970s, a question that must be considered is, "What 'demand' is there for enhanced international cooperation in this field?" After all, strongly nationalistic space programs dominated the Cold War and have operated effectively in a number of countries since then. But space is becoming more crowded and accident prone, thus putting a higher priority on international management than ever before. In the commercial sector, companies need reliability and legal enforcement mechanisms if they are going to operate profitably in a shared environment. If not for the International Telecommunications Union, for example, the world might well have seen countries seeking to seize and occupy slots in GEO, creating incentives for offensive and defensive weapons in that area of space and the likelihood of considerable environmental degradation from hazardous orbital debris. Such a situation would benefit no one. But rules are hard to create and enforce in a transnational realm like space, meaning that effective governance is a tall order.

Under the Bush administration, the United States asserted in its 2006 National Space Policy that Washington would resist any effort to restrict its "freedom of action" in space. Such language had never been used in a U.S. space policy and seemed to have been inserted by neo-conservatives in the Bush Pentagon along with another unprecedented phrase stating a U.S. policy to "oppose" any new treaties for space. The Obama administration dropped these elements from its 2010 National Space Policy. While attractive in theory, the notion of total independence in space activity is impossible in the twenty-first century. In fact, if all countries asserted their rights to freedom of action without restraint, space would certainly be ruined by orbital debris and other collective dangers. Instead, conditions of *interdependence* in Earth orbit suggest that there is a collective self-interest in the formation of clearer rules against harmful behavior, the promotion of incentives to bolster these rules, and the implementation of effective monitoring to enhance their enforcement. This means that all countries must give up some degree of freedom. In return, however, they stand to receive the benefits of restraint by other actors as well. Had the United States and the Soviet Union each not given up the right to test nuclear weapons in space in 1963, there is no doubt that both they and other countries—

such as France, China, Israel, and India—would have conducted additional space tests by now. Thus, in this case, a self-restraint treaty that included the two leading space powers worked, and it later stimulated positive, "follow-the-leaders" behavior by others. How might this lesson be applied to the problems in orbit today?

The most serious problems today have to do with destructive actions in heavily traveled regions of space (such as LEO and GEO), the risks posed by unrestricted proximity operations (when two satellites come into close quarters),[8] and the development and testing of new weapons systems (lasers, space-based jammers, and microwave systems). Countries are considering various space weapons in almost all cases not because of inherently offensive intentions, but because of fear of the activities of others. In this context, a range of possible new space security mechanisms seem to be worth investigating.

At one end of the spectrum is the simplest approach: one country declaring what it will do unilaterally to improve stability and security in space. This could take the form of a declaration rejecting the orbiting or testing of space weapons writ large, or at least specific types of weapons—such as kinetic debris-producing systems in highly trafficked regions of space. Such a declaration by a leading space power, while initially somewhat risky, could stimulate copycat behavior by others. (It could also be withdrawn if others don't follow.) If it succeeded, successive pledges by others might gradually create an international norm that only a country willing to be branded a "rogue state" would consider violating the pledge, particularly given the hazards posed by long-lasting orbital debris.

Another, slightly more complicated diplomatic mechanism for improving space security is bilateral agreements. Such pacts worked particularly well in the U.S.-Soviet context during the Cold War. Today, in a space environment with multiple actors, they are likely to have somewhat less impact. Nevertheless, countries are not uniformly equal in space, and agreements between or among leading powers (such as the United States, Russia, China, the European Space Agency, India, Israel, or Japan) would likely have significant spillover effects on other actors. For example, an effort by the United States and Russia to extend their legal restriction against interference with national technical means to other major space actors (assuming mutual acceptance of non-interference) could go a long way toward extending this norm and reducing current military space tensions.

At the far end of the options spectrum is the possibility of new international agreements involving larger numbers of actors. These include codes of conduct, conventions (such as the 2007 UN Space Debris Mitigation Guidelines), and formal treaties. Depending on the nature of the agreements, the level of domestic approval required, and the intrusiveness of verification and/or enforcement mechanisms, negotiating these arrangements can be more (or less) difficult. But formal agreements do offer real advantages in terms of reliability, stability, clarity of rules, effectiveness of implementation, and longevity. U.S.-Russian nuclear arms control efforts would have been ineffective without treaties and effective verification. Similarly, worldwide efforts to prohibit and destroy chemical weapons would not have been as successful as they have been without the 1993 Chemical Weapons Convention, its timeline for dismantlement, and its verification mechanisms. Space may have specific threatening activities that could be most effectively addressed by treaties as well. To support such efforts, it may also be beneficial to create an international space monitoring organization to supplement or work alongside existing U.S. military systems. The U.S. Air Force is already working in this direction through its cooperation with allies and even, on a more limited basis, with China, to provide warning of potential collisions involving their spacecraft. But the United States cannot be expected to pay for such an international mechanism. If countries want the benefits of enhanced transparency and the stability it could bring, they are going to have to devote adequate resources to the task, just as the United States has done out of its own national security interests in developing its Joint Space Operations Center. Private companies have also begun to pool resources in the Space Data Association (mentioned in chapter 4). It remains to be seen if the countries that are worried about space weapons tests, orbital debris, and various types of harmful interference will be able to cooperate both politically and financially to support the creation of such a system.

CURRENT SPACE GOVERNANCE PROPOSALS AND DEBATES

International debates about enhancing space security over the past decade have ranged from proposals by experts for the formation of a broad international space organization, to more specific new space treaties, to less ambitious codes of conduct, to ongoing discussions within the United Nations and the UN COPUOS, to purely voluntary measures of self-restraint. For interested parties and many observers, the progress of these efforts has seemed glacial. This point only highlights the difficulty of reaching consensus in a field where there is a great deal of dual-use technology, enduring military distrust among leading actors, and questions about the viability of future agreements in terms of compliance and verification. Nevertheless, several of the current initiatives are worth discussing in greater detail.

The most well-known and long-standing effort is that associated with an annual UN resolution on the Prevention of an Arms Race in Outer Space (PAROS). Its aim since the early 1980s has been to restart negotiations at the CD in Geneva toward a strengthened arms control treaty for space to supplement the Outer Space Treaty. Due to the lack of U.S. support (and outright U.S. opposition during the Bush administration), the resolution's effect has been mostly political. The Obama administration has thus far abstained (usually alone among the UN nations) to register its opposition to the resolution's mention of a Russo-Chinese treaty proposal it does not support on banning space-based weapons (in part because of gaps in verification). This stance has won Washington few fans on the international level. The United Nations overwhelmingly approves the PAROS resolution every year. However, it has had little effect because of the CD's failure to agree on an agenda to actually initiate the discussions on space security that the resolution demands.

Beyond PAROS, three main initiatives have received the bulk of attention in recent years. First, there is the above-mentioned initiative by China and Russia to go beyond PAROS in proposing a new treaty to ban space-based weapons, called the Treaty on the Prevention of the Placement of Weapons in Outer Space. The United States and other countries have opposed this treaty, citing a variety of specific weaknesses in the draft. Second, there is a Russian-led initiative to promote transparency and confidence-building measures in space. This effort succeeded in creating a UN-endorsed Group of Governmental Experts (GGE) that investigated the status of space security and issued a report proposing new mechanisms for improving international cooperation. Finally, there is a proposal, originally developed by the European Union, for a non-binding space code of conduct. After a 2012 endorsement from the United States, the development of this code is now under discussion as a truly international document. In addition, there are also some still incipient efforts ongoing at various levels; those will be mentioned at the end of this chapter.

Treaty on the Prevention of the Placement of Weapons in Outer Space

The first major initiative is a joint Sino-Russian proposal that emerged from a 2002 joint working paper at the CD on possible elements of a treaty on the prevention of deploying weapons in space and threatening the use of force against space objects. The motivation for this initiative stemmed from frustration with U.S. insistence on the fissile material cut-off treaty (which China initially opposed) and an effort to rally international support against the Bush administration's 2002 withdrawal from the Anti-Ballistic Missile Treaty. Both sides opposed possible U.S. deployment of space-based interceptors, viewing this development as destabilizing to space and possibly threatening their nuclear deterrents. Beijing and Moscow worked on the language and issued a formal draft "Treaty on the Prevention of the Placement of Weapons in Outer Space, [and] the Threat or Use of Force Against Outer Space Objects" (PPWT) in February 2008. Many international observers, frustrated by the Bush administration's refusal to discuss space arms control, saw the initiative as a positive effort to break the deadlock at the international level, although questions about the intentions of China's 2007 ASAT test clouded this assessment. The draft treaty defined a space weapon as "any device placed in outer space, based on any physical principle, specially produced or converted to eliminate, damage or disrupt normal functions of objects in outer space."[9] It sought to prohibit their deployment in space and required countries to agree "not to resort to the threat or use of force against outer space objects." But questions immediately arose about what the ban included. The U.S. Department of State released a set of comments and questions in August 2008 calling attention to what it identified as a number of inconsistencies and vague language requiring clarification, such as whether the testing or deployment of ground-based ASATs would be banned by the treaty. To the surprise of many international observers, the cosponsoring countries issued a formal reply in 2009 indicating that the treaty sought only to ban space-*based* weapons and would allow continued development or testing of ground-, sea-, or air-based systems (kinetic, laser, or electromagnetic), saying that such tests were not easily verifiable.[10] Critics doubted the sincerity of this rationale, given the highly transparent nature of at least kinetic tests, which release orbital debris that is easily detectable by ground-based radars. Instead, it appeared to many observers that China was attempting to retain its right to continue developing, testing, and deploying its ground-based ASAT missile interceptors, as well as lasers and jammers. Notably, neither the draft treaty nor the Sino-Russian clarification letter made any mention of international goals of debris mitigation efforts, as approved in the 2007 UN debris guidelines. The treaty also offered no specific methods for verification of space-based objects that might be carrying weapons. Prospects for international support of the agreement quickly plummeted.

Transparency and Confidence-Building Measures

While the CD process festered, Russia (joined by a number of other countries) introduced a new resolution at the United Nations in the fall of 2005, "Transparency and Confidence-Building Measures in Outer Space." The proposal's general nature and informational focus garnered widespread international support (although the United

States opposed it under the Bush administration and has abstained under President Obama). After inviting other states to provide concrete suggestions in subsequent resolutions, the 2010 version of the transparency proposal included the suggestion for the convening of a representative Group of Governmental Experts beginning in 2012 to study the issue of space transparency and come up with recommendations for the United Nations. The proposal passed in the United Nations, establishing a process that led to a formal study of space security issues on the themes of improved transparency and mutual confidence-building by a group of fifteen countries whose membership included the five permanent members of the Security Council (China, France, Russia, the United Kingdom, and the United States, or P-5), as well as ten additional states. Despite its official abstention in the UN vote on the resolution, the United States endorsed the GGE concept and participated in its sequence of meetings in July 2012, April 2012, and July 2013. In the absence of official meetings at the CD, the work of the GGE represented one of the few forums where official discussions of space security on topics beyond orbital debris had taken place since 1994. The GGE's report was issued at the United Nations in the fall of 2013. It endorsed enhanced international cooperation in such areas as disaster warning, space debris mitigation, space wealth monitoring, and the long-term sustainability of commercial space development.

Code of Conduct

In the face of the CD stalemate and the Bush administration's categorical opposition to new space treaties, the countries of the European Union (EU) sought to provide a possible path forward for international space security efforts by developing a Code of Conduct for Outer Space Activities. One of the sources for this initiative could be traced back to the United States and a nongovernmental organization called the Stimson Center. Its founding director, Michael Krepon, had earlier put together a draft code of conduct for space and had spent years pursuing this concept as an alternative to the time-consuming route of negotiating and ratifying a new treaty. The EU made amendments to the Stimson Center draft and issued its own version through the Council of the European Union in December 2008, with a revised version following in September 2010. The European-led Code of Conduct proposed that countries voluntarily pledge adherence to a set of principles aimed at promoting safe conduct in space and fostering conditions for improved space security through the adoption of a set of consensual norms—much like members of a club agreeing to behave according to certain rules. These guidelines included: non-interference with one another's spacecraft; actions to minimize chances of collision or debris release; the contribution of data on spacecraft maneuvers and any problems into a shared electronic database; willingness to consult with others in case of anticipated harmful actions; and participation in consultative meetings on implementation of the code every two years. While a number of these principles repeated or reinforced elements of the Outer Space Treaty and other prior agreements, their collection in a single document with an emphasis on taking all "adequate measures to prevent outer space from becoming an area of conflict"[11] represented a serious effort to dissuade hostile trends of the past decade.

National opinions on the code initially varied greatly. Early comments from Russia, China, and India indicated a generally negative view on two scores: (1) the code detracted attention from new treaty proposals for space, such as the PPWT; and (2) the code had been written without input from a number of space powers, particularly Russia and China, as well as developing countries. Through the end of the Bush administration and the first three years of the Obama administration, the United States remained diffident regarding the code. Washington issued periodic statements indicating general support for the "process" of forming a code, but not for the document itself. One concern for the Obama administration was opposition to the code from some Republicans in the U.S. Senate and former Bush administration officials, who argued that the "voluntary" code would restrict U.S. military options in space and should therefore be submitted for approval by the U.S. Senate (as if it were a formal treaty). Code supporter

Krepon sought to rebut this claim by pointing out that former Republican presidents have routinely signed on to such informal agreements.[12] Other critics, from both the far right and the far left, criticized the code for its non-binding character, arguing that it either would do nothing to restrict an adversary's military space programs or could easily be reversed after elections by changes in government policy. As Jeff Kueter of the conservative George C. Marshall Institute wrote, "This uncertainty may generate more, rather than less, tension in space."[13]

Nevertheless, the code represents perhaps the only path forward currently in a highly polarized international space environment. In February 2012, the Obama administration endorsed the formation of what it called an "international" code of conduct, suggesting that the European draft was a solid first step but further input was needed from both the United States and other key international actors. Since then, Russia, China, and India have all softened their stances and voiced support for the code as a means of making progress toward more formal agreements. As Russian ambassador Alexey Borodavkin stated at the Conference on Disarmament in June 2012, "We appreciate positively the draft Code of Conduct in Space proposed by the EU and are ready to participate in its finalization on a multilateral basis."[14] Two such meetings have been held (Kyiu in 2012 and Bangkok in 2013), bringing in new ideas and helping strengthen consensus in the revised document.

Other Initiatives

The one other international organization in which significant new efforts at space governance are under way is the UN COPUOS in Vienna. For the past two decades, its mandate has focused on international cooperation in peaceful space activities, since the CD is supposed to deal with arms control and security. Success in fostering agreement on debris mitigation has stimulated greater support for and participation in its annual meetings. Recent topics have included space applications, natural disaster warnings, orbital debris monitoring, and long-term sustainability. COPUOS represents one of the few active and regular forums for official space discussions, and its meetings tend to be considerably more congenial than those of the UN General Assembly or the CD, which often feature hostile and divided politics. Some observers have suggested moving space security discussions to COPUOS from the CD, but countries have resisted this idea both to protect the civil space discussions, which have worked relatively successfully in COPUOS, and to prevent the divestiture of the CD from its previously successful mission as an arms control forum.

Within broader academic debates on space governance, experts periodically renew past calls for an empowered international space organization. Detlev Wolter's exhaustive 2006 study on space law and governance for the UN Institute for Disarmament Research, for example, proposed the idea of a more comprehensive international treaty for space and its establishment of an Organization for Common Security in Outer Space (OCSO), similar to the International Atomic Energy Agency or the Comprehensive Nuclear-Test-Ban Treaty Organization.[15] Wolter's proposed treaty would limit military space activities to non-offensive, support purposes, and thus ban active defenses and destructive activities of any sort, as well as the basing of weapons in space. It would also require the destruction of existing ASAT weapons. Regarding the question of how this treaty would handle dual-use systems such as ground- or sea-based missile defenses, Wolter proposed an international system of on-site verification. In addition, the new OCSO would supervise space activities and provide early-warning information on missile launches. Finally, it would work to develop and implement rules for space conduct and develop governance mechanisms through the United Nations to implement them, possibly assimilating existing bodies like COPUOS. Despite the comprehensive nature of this proposal—or perhaps because of it—the recommended treaty and organization have yet to gain much official support from spacefaring countries.

CURRENT PROBLEMS AND POSSIBLE OPTIONS

As we have seen, the United States actively led and promoted space diplomacy in the period from 1963 to 1975. But it has resisted new space treaty efforts since the late 1990s. Given emerging problems in space, however, Washington has recently outlined a path toward a possibly more active role in space diplomacy in the future. In this regard, the 2011 U.S. National Security Space Strategy noted that "U.S. diplomatic engagements will enhance our ability to cooperate with our allies and partners and seek common ground among all space-faring nations."[16] Moreover, the 2010 U.S. National Space Policy stated that the United States will consider new treaties and legal controls if they are verifiable and serve U.S. national interests. Can the United States use its influence and its current space leadership role to "shape" the future space security environment, as it did in the 1960s when it negotiated agreements covering peaceful uses, no WMD in space, and no national appropriation of territory? If so, what is the best option in an increasingly globalized world and in a context where other spacefaring powers must be consulted and their view included?

Clear policy differences exist among the major space powers today on the aims of space diplomacy. But there is widespread agreement on the fact that the status of space governance is problematic. As Indian ambassador Sujata Mehta explained his country's viewpoint in mid-2012,

> As this global common gets more populated and crowded, and as technology develops rapidly it becomes natural to ask if the current international legal framework on outer spaces [sic] devised at the dawn of the space age more than three decades ago is adequate to address space security challenges both contemporary and future.[17]

But India does not want to see the creation of an agreement like the 1970 Treaty on the Non-Proliferation of Nuclear Weapons, which gave a special status to those countries that (unlike India) had already tested nuclear weapons before it was signed. India fears, for example, that if ASAT weapons tests are banned in space, it will be forever placed in a "second class" status in space. It therefore speaks of the need for what Ambassador Mehta called legally binding "non-discriminatory" measures. But, then, what are the consequences if all countries insist on their right to conduct harmful kinetic ASAT tests in low-Earth orbit? These are some of the dilemmas facing diplomats as they consider a new space treaty.

The current U.S. perspective on space negotiations agrees with the Indian view on the nature of the problems in space. As U.S. ambassador Laura Kennedy stated in an address before the CD in 2012, "The world is increasingly interconnected through, and increasingly dependent on space systems, but space is increasingly at risk."[18] Yet the United States, unlike in the 1960s and 1970s, now seems to shun treaties. Quoting the National Space Policy, Ambassador Kennedy noted that the United States would require that any new legal mechanisms for space must be "equitable, effectively verifiable, and enhance the national security of the United States," but added, "We have not yet seen a proposal that meets these criteria."[19] For many governments, the fact that the United States has been unwilling to offer any ideas of its own has been a source of frustration over the last decade. The Obama administration has sought to deflect opposition at home and speed progress internationally by emphasizing "the need to develop near-term, voluntary, and pragmatic transparency and confidence-building measures (TCBMs)."[20] But there are limits to such voluntary measures. Chinese ambassador Wang Qun stated his government's position in 2012 by explaining, "As voluntary measures in nature, TCBMs are not legally-binding, and they can't substitute for the negotiation of a new legally-binding instrument on outer space."[21] In an effort to bring the United States closer to their position on the PPWT, Russia and China have begun to show more willingness to compromise. As Russian ambassador Borodavkin explains this new policy, "We have already drawn ... attention ... to the fact that nothing in the Russian-Chinese draft is 'set in stone.' This is rather an invitation to a dialogue and joint creativity [sic] work than something static."[22] Whether this opening will eventually lead to productive talks, perhaps through the code discussions or follow-on meetings based on the findings of the UN GGE, remains to be seen.

A common point in the various national perspectives is a shared desire to make a currently less-than-comprehensive governance system in space more effective, thus promoting the ability of all actors to develop space and use it to further their national interests (whether in security, commerce, science, or exploration). In theory, such shared interests might eventually support the international "policing" of space, with a joint monitoring system for identifying wrongdoing and an institution possessing the tools and the decision-making capacity to enforce common space rules. But we are far from attaining such an objective, due to current mistrust and policy differences. Still, this could be kept in mind as a possible long-term goal, if international cooperation emerges in other realms and steps are taken toward greater collective security measures in space. In considering the challenges facing the maritime domain, Admiral Michael Mullin, the then chief of U.S. naval operations (and later chairman of the Joint Chiefs of Staff), enunciated a vision in 2006 of a multinational thousand-ship navy that could work cooperatively to police the seas against smuggling, piracy, and other shared threats to international security, as well as share duties in humanitarian assistance, disaster relief, and other emergencies.[23] Is this idea possible for space in the future, perhaps through the cooperation of scientists, the commercial sector, and amateur astronomers? If not, what are the alternatives? Military space dominance by one country, space anarchy, or something else? The dilemma of national enforcement in an international realm poses certain inherent problems in the absence of a dominant power. But if a leading power is unwilling or unable to carry this burden—or if the actions to achieve such a situation might actually worsen international security—perhaps the messy process of learning to cooperate for humankind's common interests is the most promising option.

CONCLUSION

The current system of space governance has emerged sporadically since 1957. Its foundations took shape particularly in the period from 1967 to 1975, when the leading spacefaring nations realized the risks of an ungoverned environment to their own future security and their ability to maintain safe access to this valuable new realm. Today, gaps in the Cold War space framework have emerged as a range of new state and non-state actors have gained access to space through the spread of technology. Conditions of increased crowding in LEO and GEO, the spread of space debris, and the finiteness of the radio frequency spectrum for satellite broadcasting have all heightened the requirement of international cooperation for the continued use and development of space. Military tensions have emerged as well to threaten stability and raise the prospects of both conflict and warfare.

These points should not cause either observers or space participants to throw up their hands and despair that nothing can be done to avoid a collision course. Efforts like the UN Space Debris Mitigation Guidelines and the newly rechristened International Code of Conduct for Outer Space Activities are steps toward greater collaboration, albeit short of clear and binding rules and procedures for future peace and stability. But the risks of failure provide a sobering incentive to work harder toward self-restraint and cooperation in this shared environment. Diplomatic tools need to be retrieved from the traditional toolbox of international space relations and new ones created to tackle emerging challenges and promote sustainability.

In light of the tasks ahead to prevent international space conflict and manage the peaceful development of space, specifically how should we go about doing this? Chapter 7 surveys the most salient problems facing the space domain across the civil, commercial, and military realms and then considers three alternative routes toward creating an improved foundation for space security.

7

TRENDS AND FUTURE OPTIONS

The future of international relations in space poses a series of questions that remain difficult to answer. It was similarly hard to predict the future of U.S. relations with the Soviet Union in space in the late 1950s or even the early 1980s; most signs pointed to possible conflict or even warfare instead of the eventual détente that developed in the early 1970s or the close cooperation that emerged after the Soviet breakup in the early 1990s. The evolving status of U.S. space relations with China presents a complex subject for analysts. But this relationship will likely go a long way toward determining the stability of space relations through at least the first half of the twenty-first century. Of course, other space actors will matter too: a changing Russia, a rising India, a more broadly active Japan, an expanding European Space Agency (ESA), fledgling but unpredictable Iran and North Korea, an increasingly capable South

Korea, an aspiring Brazil, as well as a host of other countries with a range of capabilities. Their national programs will be joined by private companies and international consortia, plus universities, international organizations, and even some wealthy individuals. All indicators suggest that space is going to be a much more complicated place as the twenty-first century moves forward.

Space technology is also going to develop innovative tools that will be more widely distributed. New fixed and mobile applications, such as roaming broadband service from space, will further increase the importance of space assets to the global economy. The possible future ability to draw on space-based energy and mineral resources could begin a process of greater material interaction between Earth and at least the nearer parts of our solar system. New exploratory and even settlement activities on the Moon and Mars, as well as in Earth-orbital space, will change space activity itself from something rare, dangerous, and exotic into something increasingly common, safe, and routine.

But military tensions in space may cloud this possibly bright and broadly beneficial future. As more countries become active in space, the chances for conflict could increase, particularly if resource conflicts on Earth become more intense. Under these conditions, space could become yet another historical venue for resource rivalry, with military spillover effects. These points raise issues concerning space governance and international relations. Many scientists make the case that the unique challenges ahead can be addressed only through collaboration. As a recent report of the international Committee on Space Research argues, "Building a new space infrastructure, transport systems, and space probes and creating a sustainable long-term space exploration program will require international cooperation."[1] Whether such international mechanisms and associated means for conflict prevention can be developed as quickly as problems emerge in space will be a critical test of existing space norms and treaties and of the political abilities of current and future national leaders.

There are, of course, various future options for governing space, from strategic control by one very powerful country (think of England on the high seas in the early 1800s) to strong and empowered international institutions (think of the original concept for the United Nations in 1945 or perhaps *Star Trek*'s United Federation of Planets). Future space governance could fall anywhere along this continuum. While many observers might postulate the superiority of international organizations, such models are effective only when they have adequate financial support and enforcement powers. Weak and irrelevant international organizations are arguably worse than none at all (witness the League of Nations in the 1920s and 1930s), because they become poor problem solvers. What will matter is the level of investment in these bodies by the leading spacefaring nations. If they are convinced that international organizations are the best route and are willing to help staff, fund, and empower these institutions, they may stand a good chance of succeeding. However, if leading nations become convinced that these bodies are irrelevant, biased, or inefficient, they may instead choose to bypass, undermine, or simply ignore these organizations and take matters into their own hands, possibly in alliance with other actors that hold similar views.

Trying to chart the future in space is a matter of identifying the key technological variables that will influence space activity and assessing the dynamics of power and influence among major actors. This final chapter does not purport to offer a determinative solution to these questions, which is impossible in human affairs. Instead, it tries to isolate those factors that we know are going to be essential for space development to proceed successfully over the next few decades and discuss incentives and disincentives for the actors that will be making the decisions to cooperate or compete with others in various fields. At a minimum, the further development of space will require some basic level of protection of the space environment, without which at least near-Earth orbital space could quickly be ruined. Because there will be multiple actors, this analysis assumes a certain level of international interaction and some future trajectory for existing space-related treaties and institutions. Finally, the discussion that follows will include considerations related to cost and space physics, which will shape decision making about missions, programs, and systems.

In order to get at these answers, we first discuss emerging trends in the "population" of space—that is, the technologies and actors. We will then analyze likely directions across civil, commercial, and military activities, and finally look at the dynamics of international space cooperation and governance. In a nutshell, given the growth of space assets and the nature of planned activities, we need to understand what factors are likely to matter in determining how space operators (nations, companies, and private individuals) will organize themselves. To do so, it is useful to compare three different predictions for space's future— based on current debates among experts. Each involves different trade-offs and choices that may be made by the leading space actors across the spectrum of possible outcomes. These futures include, from least collaborative to most inclusive: (1) military hegemony based on relative power; (2) piecemeal global engagement (or some form of "muddling through") based on consensual norms; or (3) enhanced international institutions based on new treaties and legal mechanisms. Because space outcomes will inevitably be tied to events on Earth, we must treat relevant relations among actors on Earth as an important variable in each scenario. At the same time, we should seek to evaluate the likelihood and implications of each scenario according to space-specific factors that may influence outcomes in orbit. We should also consider whether space activities and the collaborations among actors in space might begin to exert some influence—positive or negative—on international relations on Earth.

KEY TRENDS AND CHALLENGES

Global space activity continues to expand for several reasons. Some are related to practical benefits (economic and military), others involve human curiosity (science), and still others involve the political aspects of space activity (competition). Space technology has proven to be increasingly effective in stimulating global economic activity and providing a broadening range of services, especially to industries in the world's most developed countries. Banking, transportation, agriculture, city planning, and other key sectors now depend on space infrastructure to conduct their business. Modern military forces also rely on space assets and the information they provide in order to protect their populations better. More and more countries are acquiring these capabilities due to the spread of technology and the declining costs for entering the market. Many nations acquire data services without having to own actual space assets. For other actors, space meets a scientific interest by telling them more about the universe we live in and allowing them to share that information with the rest of humankind. Indeed, judging from the public response to recent missions to Mars and images from various space telescopes, the more we learn about space, the more we want to learn. Finally, there is an undeniable political and prestige-related component in the motivation of some space actors. Governments want to be involved in space so that they are seen—both at home and abroad—as active, motivated, and capable players in the modern world. Time and again, countries that had sought to refrain from developing space programs or to cut back on them for economic reasons find themselves driven to increase space spending, establish new organizations, and commit to becoming *more* active in space for competitive reasons. In fact, the expansion of interest in space has bucked recently downward global economic trends.

Evolving Space Assets and Actors

As a starting point in seeking to understand the future "population" of space, we can predict easily that the number of spacecraft in Earth orbit will probably more than double in the coming decade, rising from some 1,000 operating satellites today to well more than 2,000 by 2020, many of them small satellites. Considering the use of cubesats and other micro-satellites in low-Earth orbit, the overall number of spacecraft could be quite a bit higher. In the decade after 2020, this trend will likely accelerate as miniaturization proceeds and launcher availability increases through the expansion of platforms and reductions in cost, in part from the growth of secondary payload options for sharing launch services. There is likely to be a similar but less dramatic increase in the number of satellites in medium-Earth orbit because of decisions by several countries to deploy their own constellations of precision navigation and timing systems. In geostationary orbit, there will also be a moderate increase over the next decade as countries seek access to critical communications, broadcast, and observation slots and as technologies able to parse the radio spectrum more finely allow the bunching of multiple satellites in the limited locations available.

Besides robotic satellites, the population of space over the next two decades will involve an increasing number of people and space stations in near-Earth orbits, for both scientific and recreational purposes. Besides the huge *International Space Station* (*ISS*), there is likely to be a multi-module Chinese space station, plus a handful of private orbital hotels and perhaps some industrial production facilities by 2020. This number could increase to several dozens or more by 2030. There will also soon be hundreds of people each year venturing into suborbital space for short periods of time, thanks to tourist services like Virgin Galactic, Bigelow Aerospace, and XCOR Aerospace. This number will certainly expand rapidly after 2020, once reliable services are established and prices drop.

In terms of the celestial bodies, it seems unlikely that more than one or two missions will return people to the Moon before 2020. But it is quite likely that multiple missions to the Moon will take place before 2030, and even that semi-permanent settlements might be established and maintained, much like the *ISS* in low-Earth orbit today. The most likely inhabitants (as on Antarctica) will be small groups of scientists (probably from a number of different countries). There may also be brief missions to one or more asteroids, including some missions involving commercial entities. Beyond that, Mars will likely be the next destination, but (for humans) not until closer to 2030 or even later due to its distance from Earth.

Finally, in terms of solar system research or missions to deep space, there will be no shortage of destinations. Such missions will increase in frequency and number, but probably still remain in the low single digits per year, given the high costs and the still relatively few countries, companies, and international consortia capable of planning, developing, funding, launching, and operating such complex missions. For at least the next decade, only the United States, ESA, Russia, Japan, China, and India will be in this league of leading exploratory countries with independent launch capability and the resources to carry out long-duration missions. Well-funded private scientific or commercial ventures (funded by speculators, enthusiasts, or advertisers) might join this group, using technologies developed by one of the advanced space programs or available from private companies (like SpaceX, Boeing, Lockheed Martin, Khrunichev, Energiya, or others).

There will also be changes in who will be operating satellites and manned spacecraft. Overall, the center of gravity among critical actors in space is likely to shift southward on the globe as the growing markets of Africa, Latin America, and Southeast Asia become a more significant part of the "demand" for space services. The number of launch sites will grow steadily as well, with suborbital flights being offered from locations in the Middle East and wherever else there are wealthy clients and a favorable regulatory climate. New orbital launch sites may emerge along the equator in Brazil, Indonesia, and perhaps locations in Africa as well (possibly funded by outside players). Among nations, the United States will probably remain the leading spacefaring country at least through 2030, although it will have to step up its current rate of investment if it wants to stay ahead of China. Russia will be challenged to maintain its current place in space expenditures[2] and will very likely fall behind ESA and China, and possibly Japan as well, unless it begins to innovate and implement a more effective strategy for reviving space research and development. As new launchers enter the market, Russia will find its market share challenged and its current lock on human spaceflight threatened well before 2020.

Significant growth will also occur in the number of companies, universities, governments, and international organizations that can afford satellites as entry costs decrease, particularly in relative terms. Space businesses will likely become more genuinely commercial, with subsidies declining in major countries and competition increasing along with consumer demand. The traditional model of state-led investment in space will not be fully replaced within twenty years, but a growing number of companies will begin to invest in space platforms with an eye to beating the market for new services. For example, the formation of companies aimed at mining asteroids provides evidence of the entrance of significant new investors into the commercial space sector. Many of these efforts are likely to be multilateral joint ventures and involve increasingly integrated combinations of international technologies, marketing activities, and financial instruments. Indeed, countries whose national companies are forced to go it alone will be at a significant disadvantage.

Nongovernmental institutions will be joined by wealthy individuals who will become more directly involved in space activity through personal travel and related investment behavior. Over time, this pool of individuals will grow into the thousands, which will likely stir greater public and media interest in space services, infrastructure, and profits, while providing new opportunities for private space exploration.

TECHNOLOGICAL TRENDS AND TRADE-OFFS

Space activity over the next two decades will at some level be an expansion of prior and existing activities. But at another level, new areas of focus and fundamentally new missions will emerge that will change the center of attention and shift some of the critical areas of interaction. As a first cut, space activity in the next decade will continue a process of infrastructure building in areas such as space communications, remote sensing, precision navigation, and transportation. Services will become broader, more sophisticated, faster, and more reliable. But there will also be areas of expansion, such as direct satellite broadband service for mobile users, more tailored mapping and imagery data, and new satellite access via smaller receivers and voice recognition software. More important, there will likely be more experiments during the coming decade—for example, hardware demonstrations of laser communications systems, which could eventually revolutionize the industry by freeing providers from dependence on an increasingly finite radio spectrum. Finally, in the energy sector, the next ten years will likely see the establishment of orbital refueling for satellites and at least prototype delivery of space solar power. Within twenty years, this technology could free spacecraft of the need to carry heavy solar arrays (thanks to energy provision by space-based solar farms). Eventually, space solar power could also provide a portion of Earth's energy needs, although full-scale operation of such an infrastructure will likely have to wait for a reduction in launch and in-space construction costs as well as for the highly accurate beam-focusing technology required to deliver energy from orbit to Earth safely.

As noted above, space tourism will become a much more robust industry by 2020 and especially in the decade after that, when such services might become accessible to those who fall below the top one percent of earners in the developed world. Indeed, much like air travel after World War II, it is foreseeable that suborbital flight will become affordable to tens of thousands of people in upper-income brackets by 2030, with a range of new services available as technology develops further. Hundreds may be able to visit orbital hotels or stations within ten years, and a growing number of people will be working in space, tending tourist facilities as well as various industrial and manufacturing enterprises.

Another factor that might change the direction of current activities in Earth orbit is the expansion of national military programs in space. To date, only three countries carry out significant military activities beyond reconnaissance: the United States, Russia, and China. But this group will likely expand further in the coming two decades. The list of militaries that might decide they have a strategic interest in testing kinetic, laser, or other active space defenses includes India, Pakistan, Japan, Iran, Israel, France, Brazil, and North Korea. Additional types of weapons that might be developed and tested against space objects in the coming two decades by various militaries—assuming new arms control mechanisms are not developed—include microwave systems, particle beams, space mines, and Earth-, sea-, air-, and space-based electronic jammers.

At present, no treaty forbids these technologies, and there are strong military-industrial lobbies in a number of countries supporting space-based weapons, despite their possibly disruptive effects on space commerce, science, and passive military operations in the same regions of space. In all likelihood, the growing population in the lower reaches of space will force some sort of decision regarding priorities:

either to allow countries to test and deploy large-scale orbital defenses or to strictly limit destructive weapons and emphasize commercial development of low-Earth orbit, including expanded human space-flight. Active defenses and commerce probably will not be compatible in crowded orbits because of the linkage between space weapons and harmful debris, particularly since such military tests and related on-orbit deployments—once begun by one country for missile defense, ASAT, or space-to-ground attack options—are likely to be met with countermeasures by other militaries. Under such conditions, the development of commercial human spaceflight in low-Earth orbit will become too unsafe to continue.

In this regard, successful space traffic management will be essential to the ability of people, companies, and countries to enjoy future services. This improved policing must include preventing orbital collisions with debris and other spacecraft as well as avoiding radio frequency conflicts. To date, success has arguably been possible less because of effective international management and more because of the lack of crowding in space. These conditions will no longer hold in the future.

Given these challenges, it seems unlikely that current space governance mechanisms will be adequate to the task of managing foreseeable space risks across the range of new actors and activities. For this reason, we next look at three alternative mechanisms for managing space over the next two decades: military hegemony, piecemeal global engagement, and enhanced international institutions.

THREE ALTERNATIVE SPACE FUTURES

International order has historically depended on the actions of a few powerful countries, which have made rules and imposed them upon others. Depending on the international balance of power at the specific time and the geographic breadth of the required reach of the governance mechanism, this could be accomplished by a single hegemon or by an agreement of a few like-minded actors. The Bretton Woods system and the roots of the Western capitalist economic system after World War II owed much to U.S.-created rules and institutions, supported by American funding. Similarly, these Western structures relied on a system of security provided by the U.S. military. Internationally, a later a network of agreements with the Soviet Union that emerged after the Cuban Missile Crisis of 1962 helped stabilize security relations. These included the Partial Test Ban Treaty of 1963, the Treaty on the Non-Proliferation of Nuclear Weapons of 1970, and a series of bilateral nuclear arms control treaties from 1972 to 1991. In space, U.S.-Soviet leadership proved critical to the formation of the 1967 Outer Space Treaty, the now-defunct Anti-Ballistic Missile Treaty, and the series of UN conventions on liability, registration, and broadcasting (discussed in chapter 6).

But leading international actors since then have failed to make further progress toward strengthened collective security in space. This is a problem because emerging conditions pose challenges not anticipated by the limited Cold War–era agreements. Yet the international community may also have new mechanisms to manage space that did not exist until recently, thanks to technology. On the minus side, national space programs are not the only actors anymore, and traditional mechanisms to enforce these agreements are proving inadequate in a world where the United States and Russia can no longer put overwhelming pressure on an offending state within its sphere of influence. But, on the plus side, new capabilities and data-sharing options (such as the Internet) mean that policing space by a group of space powers or an engaged space community might be possible and even more effective than the old tools. With these points in mind, it seems clear that new means of maintaining safe access to space need to be investigated. The question is, "How might they emerge, if at all?" This section looks at three alternative predictions and discusses their likelihood, costs, and benefits.

Military Hegemony

One possible route for managing space and preventing future conflicts over scarce near-Earth resources might be the establishment of space dominance (or hegemony) by a single powerful country. Those who predict such a future, such as the space strategist Everett Dolman, refer to historical struggles on newly discovered continents (such as Africa, Latin America, or the New World) or domains (such as sea and air). If such a struggle occurs over space, they see the United States as the best candidate to play the role of hegemon.[3] It currently has the largest space program by a significant margin, and its democratic-capitalist orientation will ensure that all law-abiding actors can participate in space's development without risk of harm by an aggressive or authoritarian adversary. This perspective views the post–Cold War state of affairs in space—where multiple parties are vying for power and influence and enforcement of rules is difficult due to the lack of a single authority—as inherently unstable, suggesting that there will be an eventual winner and an array of losers in a struggle for ultimate space power. As Dolman and former Reagan administration official Henry F. Cooper, Jr., make this case, "With great power comes great responsibility. If the United States deploys and uses its military space force in concert with allies and friends to maintain effective control of space in a way that is perceived as tough, nonarbitrary, and efficient, adversaries would be discouraged from fielding opposing systems."[4] For this reason, they favor a proactive policy by the United States to seize control over low-Earth orbit and to deny access to any country that might threaten to disrupt space through harmful weapons development or other actions. Dolman estimates that this effort would require seeding the lower reaches of space with offensive weapons in order to police the global launch environment, costing an estimated $3–$5 trillion.[5] Otherwise, he anticipates a similar effort by China or a resurgent Russia, with far less favorable results. In today's Washington budget environment, the prospects here seem dim. Fortunately, even a growing China might find the price tag too expensive.

However, if we extrapolate from traditional power-based arguments and relax this model's assumption of a single national actor, the argument becomes somewhat more realistic. In fact, the rise of various friendly space powers in recent years suggests that it may not be necessary that the United States or China—whose resources are likely to be too limited to accomplish such a task alone—would be the only option as a space hegemon. Instead, a coalition of like-minded countries linked by national military and commercial interests could join forces to share the burden and increase their effectiveness. One could envision perhaps a U.S.-European-Japanese-Indian space alliance or perhaps a Chinese-Russian-Iranian one. If a range of launch sites were linked across the globe and the use of networked space resources and defenses were shared, the concept of space hegemony seems somewhat more realistic, as well as affordable.[6]

Yet any scheme based on space dominance must overcome a key obstacle—without hegemony by a single country or a coalition of countries in the balance of power on Earth, it seems implausible to think that space control could be accomplished separately without serious opposition. Thus, as any leading nation or coalition sought to deploy its space weapons for whatever purpose, other powerful countries would obviously seek to counter this hegemony by launching their own. The space analyst Nancy Gallagher summarizes cogently: "The United States cannot unilaterally protect all of its satellites, or prevent others from acquiring the means to threaten them, even if it dramatically increased military spending."[7] Dolman and Cooper assert the possibility of "cowing" rivals into submission. When viewed through the lens of history, however, this seems an unlikely outcome and one replete with risks, among them the possible ruination of space by debris from kinetic encounters as one side or the other seeks to test and deploy systems intended to cement their individual or collective dominance. It also discounts the possibly important moderating role of new actors, such as private companies and international consortia, which are becoming increasingly influential in affecting space outcomes and are more interested in profits than in power-driven warfare in space. For these reasons, old-style hegemonic theories for space—drawn largely from military concepts of previous centuries—seem unrealistic. They also seem highly unlikely to lead to beneficial and sustainable outcomes either for the United States or for other space-faring nations and users.

Piecemeal Global Engagement

A second possibility in current debates on future space management is the one posited by the current system of limited, piecemeal treaties and ad hoc problem-solving. This path of muddling through has arguably worked so far and could conceivably be extended into the future through new informal agreements and the gradual expansion of space norms, rather than the establishment of a single military hegemon or, alternatively, new international space institutions and treaties to govern space. Instead of assuming the continued exclusive role of national governments, this approach is flexible enough to include a variety of other actors, who could perhaps manage and prevent conflicts more effectively through the use of market-based mechanisms or other informal rules. This model draws on the concept of "soft governance," or the idea that international behavior can be managed by developing new roles and expectations that pressure actors in a community to behave according to consensual social norms, much as people do in their daily lives.[8]

Today, initiatives like the International Code of Conduct for Outer Space Activities embody this approach. Given the growing political difficulties of treaty ratification (especially in the United States) and the limits of strict enforcement, the code could in the future work through a self-policing mechanism and a kind of "neighborhood watch" system, wherein information would be shared and countries would be "blamed and shamed" by exposure of their harmful activities to the broader community. This would lead to sanctions that would hurt the violator's economic standing in space and on Earth, resulting in the loss of soft power and influence, with eventual military and security implications as countries began to rally against the violator.

In the commercial realm, evidence of such soft governance can be seen in the Space Data Association, which now includes numerous private companies and a few governmental organizations that have chosen to provide and share data on satellite locations and maneuvers to reduce risks of collision. This kind of flexible, ad hoc management may be the wave of the future in space. According to the Space Foundation, such innovative "space community partnerships" may be formed naturally because of "the ever-growing technical capabilities distributed throughout the government, industry, and academic sectors."[9]

In the case of the approaching second phase of lunar exploration, it remains unclear who will organize the next major effort to put people on the Moon and to develop a possible base there. While in the past one could easily make the assumption that it would be a major state or, more recently, a combination of countries (following the *ISS* model), the expansion of space actors in the twenty-first century offers other options. Two space analysts predict a new era that they call Space Exploration 3.0: "This new phase of space exploration results from an organic evolution and will, unlike the previous two space exploration eras . . . not only include countries through their space agencies, but also industries, universities, research institutions, and other non-governmental organizations."[10] Such hybrid efforts may require more flexible governance models than even the government-led *ISS or* the recent International Space Exploration Coordination Group (ISECG). Here, organizations of independent scientists like the International Lunar Network could play a role or possible commercial organizations might emerge organically from shared interests in lunar infrastructure (such as for housing, fuel, and food). International organizations like the UN Committee on the Peaceful Uses of Outer Space, which has become an increasingly active forum in space sustainability discussions, could play a role in developing guidelines for exploration, settlement, and cooperation.

These informal and tailored governance structures for space might offer certain benefits, but they suffer from drawbacks as well. Specifically, by failing to include considerations of national power and by largely excluding military considerations, they risk the emergence of serious security-related problems that such soft organizations are incapable of handling. At the same time, enforcement of sanctions against violators of norms is difficult without the kind of clear legal rules and formal mechanisms that reside in national governments or empowered international institutions. Thus, while these approaches provide useful paths forward in areas like space commerce and scientific exploration, they seem less well suited to grapple with the major emerging security-related problems on the horizon in relations among the leading space powers: the United States, Russia, China, India, Japan, and others. Even if new, voluntary organizations or crowdsourcing are successful in their specific efforts, space could be ruined by unregulated activities in the military arena, over which the voluntary entities would likely have little influence or jurisdiction. The emergence of a tacit U.S.-Russian norm against kinetic ASAT testing by the 1990s failed to prevent China's ASAT test in 2007, due to Washington's and Moscow's failure to establish this norm legally via an international treaty. Similarly, an initiative like the International Code of Conduct, which does involve multiple national governments, is still only a voluntary effort that provides at best a first step toward new norms of behavior, rather than addressing gaps in existing treaties and threats to existing space stability in a direct and binding manner. For this reason, other experts believe that more-formal arrangements and empowered international groups or organizations may offer more significant benefits for space.

Enhanced International Institutions

If space travel is to become more like air travel, it will require a set of coordinated national regulations to ensure safety and set standards for acceptable behavior. It will also have to be backed by a legal regime able to protect businesses and the public alike. To date, the limited scale of space tourism has not merited the effort required to accomplish this task. It seems very likely, however, that demand will grow in the coming decade and cross that threshold. Both businesses and governments will want clearer rules and procedures, even if the two sides disagree on the proper balance between freedom to operate and protection of both passengers and people on the ground.

Similarly, given the limited number of locations in geostationary orbit above the equator and the finite radio frequency spectrum, there seems to be no viable alternative (short of some hegemonically imposed order) to a collaborative governance model for satellites. This framework has existed since the 1970s, but, as discussed earlier, it has become increasingly stressed by the rapid growth in demand. Space traffic control is a related problem that is only going to become a more serious priority in the future, and it is an inherently *international* problem.

Space security is currently governed by a network of limited, ad hoc treaties formed during the Cold War under the tutelage of the two superpowers to address specific problems. Power mattered for bringing this framework into existence and maintaining it during that period. Today, the rise of new actors and activities in the context of unresolved security problems in various regions populated by hostile military rivals presages the possible exploitation of gray areas and outright gaps in these agreements. As Gallagher predicts, achieving space security in the future will "require formal negotiations, legally binding agreements, and implementing organizations that have both resources and political clout."[11] The worsening problem of space debris and the playing out of regional and, in the U.S.-Chinese context, global space rivalries represent the most threatening issues for the future of space security.

The German space analyst Detlev Wolter, reflecting on the absence of space conflict since 1957 and the recent emergence of new military space tensions, observes:

> It would be an irreparable setback for the international community to now lose the peaceful purpose standard in outer space, and risk having space become the new arena for an arms race for the sake of unilateral military "space control" ambitions and the transgression [*sic*] towards active military uses of a destructive nature. If met successfully, the challenge will inspire mankind's hope that the common space will be governed by an internationally agreed upon *pax cosmica*.[12]

Notably, among U.S. military officers there is a growing recognition of the risks of ruining space and of the need to enlist all countries in focusing on long-term preservation and debris mitigation. Colonel (U.S. Army, Retired) Patrick Frakes sees the need to establish "an organization representing all nations conducting space activities which has the technical competence and commitment of members to optimize current and planned space activities with the goal of sustaining the environment for future generations."[13] This is challenging, given current international space competition. Will countries be willing to sublimate their desire for unilateral advantages to the broader interests of humankind in safe access to space and peaceful commercial development?

At present, arguably the most viable model for such cooperation is the *ISS*. While no one familiar with the organization's troubles would argue that it represents an unqualified success, it does offer an example of a formula that has worked and held together with remarkably few problems besides the normal delays and financial squabbling inherent in any large international activity involving sophisticated technology. The *ISS* originated through hegemonic (U.S.) leadership at a time of shifting international order (the end of the Cold War) and has a lopsided funding structure largely dependent on the United States. Nevertheless, it has successfully integrated a number of previously distrustful actors and created a space-based research module of impressive scale. The *ISS* structure is not an ad hoc norm-based mechanism, but instead a formal organization based on intergovernmental agreements that specify a complex formula of contributions and access for each country. Its success has stemmed from the clarity of its rules and the fact that it is a limited self-interested body not dependent on the United Nations or large numbers of spacefaring countries. It is unlikely that a less formal agreement would have worked. As we look to return humans to the Moon and eventually send them to Mars, the *ISS* model has a number of advantages: a legal structure, joint funding, coordinated (but not jointly produced) technology, and a mutually beneficial division of labor. Looking back, the station could not have been built without the U.S. space shuttle and Canada's manipulator arm. But without Russia, no country would have had access during the shuttle's down periods or during the current gap in U.S. human-rated launchers. Finally, there would not be much to do on the *ISS* without the research modules contributed by Japan and ESA and the equipment and supplies brought up by their cargo vehicles.

The U.S. Constellation program had some of the same characteristics as the *ISS* model. Though some partners complained that the draft plan offered no "critical path" technologies to the non-U.S. players, it would probably have worked had Congress provided and the public supported adequate funding. Whatever is decided in the future about lunar and Martian exploration and settlement, it seems likely that multiple governments (and possibly some private corporations) will have to be involved in order to share the massive costs. These requirements could make a formal arrangement more beneficial than an informal one. Moreover, as exploratory activities evolve into possible resource extraction, such a structure could be highly beneficial to encourage coordination and reduce conflicts. The legal regime for commercial activities on the celestial bodies contained in the Outer Space Treaty is clearly inadequate to manage foreseeable activities within the next twenty years without further intergovernmental elaboration and agreement.

In the space security field, the major spacefaring nations have thus far been unable to identify areas of consensual agreement for new treaties. Formal agreements are not in and of themselves to be desired, but the history of space (with regard to nuclear testing, for example) and of other environments suggests that treaties—if verifiable—can be the best means of preventing harmful activities, particularly if they are put in place before such capabilities are widely distributed and systems tested and deployed. As seen with the proliferation of nuclear weapons, it is much harder to "undo" weapons once they have been fielded and become international symbols of status and power.

For this reason, the most logical future progression may be to begin with a voluntary code of conduct but then use the consultation mechanism built into it to identify areas of further concern that may benefit from a more explicit and legally binding agreement. Such initiatives might include collective efforts to rule out interference with space-based global utilities such as the GPS system, the creation of an international space situational awareness (SSA) body to prevent collisions, and special prohibition of kinetic activities against space objects in any region of Earth orbital space where long-lasting debris will be generated. This should include all regions of near-Earth space from geostationary orbit to around 150 miles above Earth, below which these actions would have to be sharply limited and announced well in advance.[14] Such a formal regime could go a long way toward promoting stability and could be verified by a cooperative SSA organization. Banning destructive testing of any sort using lasers or other directed-energy weapons, microwaves, or space mines would strengthen such a security structure and could be verified by both ground-based and possible new space-based sensors.

However, the formation of such an enhanced space security regime would require close collaboration among the major spacefaring nations. The leading space-launch countries and their militaries would have to decide that such a legally binding framework is in their interests. To date, the United States and China have had only very limited high-level discussions on space security, and the Russian and U.S. governments remain divided over missile defense questions and challenges in nuclear weapons reductions. India has offered rhetorical support to prohibitions on space weapons but few specifics thus far, and wants an anti-satellite capability to counter China. Europe, Japan, and South Korea seem amenable to further limits if the United States goes along and they are not locked into a discriminatory regime. Iran, Israel, and North Korea are wild cards, but thus far lack any apparent space weapons. Thus, at present, a universal regime that allowed only testing against suborbital objects (such as ballistic missiles) at low altitude could lock in protections in higher orbits and not be discriminatory. Fortunately, as noted earlier, the growing presence of space tourists in low-Earth orbit would greatly increase the incentives for restraint in any future test programs. Such inducements could also be addressed through supporting bilateral or multilateral efforts and possible civil space benefits or commercial privileges distributed by the leading spacefaring nations. Such a set of agreements in space security could work and would serve the self-interests of all major actors.

That said, the leading spacefaring nations have a long way to go if they are to pursue this route successfully. Discussion on the International Code of Conduct and follow-up activities drawing on the report by the UN Group of Governmental Experts offer prospects for vetting these concepts in an intergovernmental context to see if they are viable. Similarly, the steady efforts of scientists and companies to promote collaboration and share information on common threats in space (such as the Space Data Association) could smooth the way for innovative mechanisms at the nongovernmental level. In this manner, ad hoc and informal efforts to improve flexible space governance tools could foster more-formal arrangements later on as areas for critical state-level agreements are identified and consensus is built. The alternative may be the dangerous disintegration of existing space agreements, including the 1967 Outer Space Treaty, under competitive military pressures from existing and emerging space actors and from the limits of purely voluntary mechanisms alone to create meaningful collaborative restraints on harmful space behavior. Unfortunately, such an outcome is not hard to imagine. But there would be considerable costs to the existing and future space economy, the military support networks that benefit myriad countries, and the prospects for the peaceful scientific discoveries, material resource development, and energy generation that will improve future human life on Earth—and in space. Whether fear of the loss of these benefits will be a powerful enough deterrent to generate progress in the form of new international agreements regarding space remains to be seen.

CONCLUSION

The overriding question of whether military or commercial/scientific activities will dominate the future of international relations in space still hinges to a considerable degree on whether or not countries succeed in preventing major conflicts on Earth. But significant military confrontations on the ground, sea, and air would almost certainly spill over into space. Fortunately, forces of economic globalization mitigate the chances of war among today's great powers, but they have not by any means eliminated them.

Short of such conflict, the challenge of space governance is in some ways akin to solving environmental problems on Earth, including global warming and air and water pollution. We have learned a great deal about these shared risks, but self-interest and the growing fragmentation of power among countries, groups, and organizations make it difficult to reach a consensus and translate that into action. However, given the fragility and limited debris-carrying capacity of near-Earth space, a catastrophic tipping point may be closer in orbit than it is on Earth. Indeed, collaboration may be the *only* way of maintaining sustainability in space. Moreover, unlike the situation with environmental problems on Earth, national militaries have perhaps the greatest interest in protecting the *space* environment, given their disproportionate investment in existing assets.

Despite periodic flurries of interest by the Soviet, U.S., and most recently Chinese militaries in space dominance schemes, there is an increasing realization that acquiring such capabilities would undoubtedly involve such enormous costs and strong foreign reactions that they would never be worth it. Such deployments would inevitably lead to conflict and the possible ruination of near-Earth orbital bands. Instead, countries—for better or for worse—need to get down to the difficult business of learning how to work with one another and to use space for what is really most valuable—*information* and, at some point, energy and other resources. Indeed, if any lesson is to be learned from the experience of military activities in space since 1957, it is that space is too valuable to risk ruining it for the sake of short-term military objectives. At best, temporary jamming of signals going to and from space represents the only sustainable means of "fighting" in orbit, if countries want to continue to enjoy safe access to this fragile environment.

In terms of commerce, space remains a relatively untapped region. Space-based assets bounce signals from one place on Earth to another, broadcast information to large areas on the ground, and record various types of images of man-made objects and the geographical features of the land and the seas. Space-derived assets—such as space-based solar energy and minerals on the celestial bodies—have yet to be developed. While some environmentalists seek to "protect" space from such exploitation, this process seems inevitable, given our needs on Earth and in space. The real challenge will be how to develop and distribute these resources in a manner that meets three objectives: (1) allowing entrepreneurs to profit from the necessary new technologies, (2) the protection of the space environment, and (3) the prevention of international conflict over the resources. Challenges also exist, and are intensifying, in the management of near-Earth orbital resources affecting satellite locations and broadcasting. To date, international mechanisms have mostly worked, but they are experiencing pressure from developing countries seeking a greater role and from authoritarian states seeking to isolate their populations from satellite-delivered information. Initiatives with the United Nations and within the business community are beginning to tackle these problems. But sustained political efforts will be needed and international regulatory and enforcement breakthroughs will have to be made to prevent national attempts to turn to military solutions.

On the Moon and other celestial bodies, much more specific rules will be needed to manage semi-permanent research stations and possible commercial mining activities. Although free marketeers call for abandoning the Outer Space Treaty, privatization of ccelestial bodies, and a free-for-all, to the benefit of the early innovators, such processes would have too many negative effects to be worthwhile. But rules for harmonizing the activities of innovators need not be unduly onerous or bring any requirement for international control over profits. What is needed are judicious discussions among the leading spacefaring nations to determine how best to coordinate new activities without causing conflict. The United States and the Soviet Union took the first major step in agreeing not to claim territory or engage in military activities before they reached the Moon in the late 1960s. The next group of explorers, developers, and settlers should decide—perhaps in consultation with UN bodies like COPUOS or space agency bodies like the ISECG—how to pursue their interests while not ruining the Moon or denying later participation to other parties. In the end, it is possible (and even likely) that shared interests in survival and in creating a pool of common services may nudge rival efforts toward cooperation in this harsh environment, as has been seen in Antarctica. The major difference between the Moon and Antarctica, which has been largely preserved as an international research continent, is that commercial activities are both allowed and expected on the Moon. How these initiatives will be cooperatively managed thus represents the biggest single challenge.

Space exploration continues to fascinate people on Earth because it helps us learn more about the universe and offers options for preserving human life beyond our planet. As technological advancements in remote-sensing, long-distance communications, and spacecraft transportation allow humans to reach beyond near-Earth space, into the solar system, and past our galaxy, we are likely to uncover truths that will expand our understanding of the meaning of the universe and the opportunities offered by exploring it. In doing so, we may also be able to use scientific information to help break down barriers between countries and peoples on Earth, particularly if other life-forms are eventually found in the universe. In this way, space could gradually have a positive effect on human relations on Earth.

Early space visionary Arthur C. Clarke wrote, "Only through spaceflight can Mankind find a permanent outlet for its aggressive and pioneering interests."[15] But space has not yet had this effect. Only when countries recognize the risks of the permanent extinction of our species from either man-made problems or from the impact of dangerous space objects colliding with Earth are we likely to begin meaningful cooperation in protecting Earth and near-Earth space, collaborating in long-duration space exploration, and reducing the kind of nationalism that has stimulated harmful military activities in space. There is some evidence of this behavior through the *ISS* experience, but we have a long way to go.

Fortunately, human beings are purposeful creatures. We are also capable of broadening our minds and using new information to alter our behavior. But it is hardly reassuring that although we created nuclear arms, we have not yet killed ourselves in a cataclysmic nuclear war. Similarly, simply knowing that cooperation in space is more mutually beneficial over the long run than conflict does not make it inevitable. Self-serving individuals and nations could ruin space's potential for humankind. Existing international guidelines are still vague in defining specific harmful activities—such as debris-creating weapons tests. The more that countries can do to strengthen reasonable norms and transform them into verifiable and enforceable rules, supported by scientists and an informed civil society, the more hopeful space's international future will become. It will be much harder to halt these technologies and associated harmful behaviors after there is extensive testing and multiple countries possess them. Such a course will require greater military leadership and more new initiatives than have been seen in the past regarding space. This point does not encourage optimism, given the gridlock apparent in Washington today and the hostile dyadic relations seen in the Sino-U.S., Sino-Indian, and even aspects of the current U.S.-Russian relationship regarding space.

In today's multinational space environment, there has too often been a tendency to think that improving security in space is somebody else's job and that things will "work themselves out" if they are simply left alone. This is wishful thinking. Instead, all spacefaring countries, companies, scientists, and interested individuals need to actively engage with the problem of conflict prevention. Only in so doing can we accomplish what is in humankind's best interests: international cooperation in the sustainable use and development of outer space.

NOTES

INTRODUCTION

1. "Star Wars: The Force Unleashed II," WebDoc #4, http://starwars.com/watch/behind-the-scenes/2.

1. GETTING INTO ORBIT

1. Willard E. Wilks, *The New Wilderness: What We Know About Space* (New York: David McKay Company, 1963), 45.
2. For a thorough history of the V-2, see Michael J. Neufeld, *The Rocket and the Reich: Peenemunde and the Coming of the Ballistic Missile Era* (Cambridge, MA: Harvard University Press, 1995).
3. Walter A. McDougall, *The Heavens and the Earth: A Political History of the Space Age* (New York: Basic Books, 1985), 6.
4. For a more thorough discussion of the dual-use nature of space technology, see Joan Johnson-Freese, *Space As a Strategic Asset* (New York: Columbia University Press, 2007), esp. chapter 2, "The Conundrum of Dual-Use Technology," 27–50.
5. This section draws on the following books: Sean Carroll, *From Eternity to Here: The Quest for the Ultimate Theory of Time* (New York: Dutton, 2010); Jerry Jon Sellers (with contributions by Williams J. Astore, Robert B. Giffen, and Wiley J. Larson), *Understanding Space: An Introduction to Astronautics*, rev. 2nd ed. (Boston: McGraw-Hill, 2004); William E. Burrows, *This New Ocean: The History of the First Space Age* (New York: Random House, 1998); Kenneth Gatland, *The Illustrated Encyclopedia of Space Technology*, 2nd ed. (New York: Orion Books, 1989); Bernard Lovell, *Man's Relation to the Universe* (San Francisco: W. H. Freeman, 1975).
6. On this history, see Frank H. Winter, "The Genesis of the Rocket in China and Its Spread to the East and West," in A. Ingemar Skoog, ed., *History of Rocketry and Astronautics*, vol. 10 of *Proceedings of the Twelfth, Thirteenth, Fourteenth History Symposia of the International Academy of Astronautics* (San Diego, CA: American Astronautical Society, 1990).

7. For more details on Tsiolkovsky's contributions to rocketry, see William J. Walter, *Space Age* (New York: Random House, 1992), 5–7; also Asif A. Siddiqi, *Sputnik and the Soviet Space Challenge* (Gainesville: University of Florida Press, 2003), 1–2.

8. On Goddard's struggles, see Burrows, *This New Ocean*, 46–53.

9. Neufeld, *The Rocket and the Reich*, 273.

10. McDougall, *The Heavens and the Earth*, 123.

11. For more on the so-called mass ratio of a launch vehicle, see Wilks, *The New Wilderness*, 48–52.

12. For more on the early history of the Soviet rocket program, see Siddiqi, *Sputnik and the Soviet Space Challenge*, 23–84.-

13. Neil Sheehan, *A Fiery Peace in a Cold War: Bernard Schriever and the Ultimate Weapon* (New York: Random House, 2009), 364.

14. On its design, see Siddiqi, *Sputnik and the Soviet Space Challenge*, 483–87.

15. For a thorough discussion of the various types of rocket fuel, see Sellers et al., *Understanding Space*, 564–69.

16. Ibid., 566.

17. Wyn Q. Bowen, "Report: Brazil's Accession to the MTCR," *Nonproliferation Review* 3 (Spring–Summer 1996): 86–91.

18. On this incident, see Siddiqi, *Sputnik and the Soviet Space Challenge*, 239–40.

19. Raimundo Garrone, "Accident Wounds Brazil's Space Program," Reuters, August 26, 2003.

20. For more specifics, see David Wright, Laura Grego, and Lisbeth Gronlund, *The Physics of Space Security: A Reference Manual* (Cambridge, MA: American Academy of Arts and Sciences, 2005), section 10, "Elements of a Satellite System."

21. As of November 30, 2012, the figure was 1,046. See the Union of Concerned Scientists' Satellite Database, http://www.ucsusa.org/nuclear_weapons_and _global_security/space_weapons/technical_issues/ucs-satellite-database.html.

22. For more on satellite orbits, see Wright, Grego, and Gronlund, *The Physics of Space Security*, sections 4 ("The Basics of Satellite Orbits") and 5 ("Types of Orbits, or Why Satellites Are Where They Are").

23. Union of Concerned Scientists' Satellite Database, http://www.ucsusa.org/ nuclear_weapons_and_global_security/space_weapons/technical_issues/ ucs-satellite-database.html.

24. Ibid.

25. Ibid.

26. Ibid., 49.

27. For a wealth of information on orbital debris, see the website of NASA's Orbital Debris Program Office, Johnson Space Center, http://orbitaldebris.jsc .nasa.gov/.

28. Nicholas Johnson, *Soviet Military Strategy in Space* (London: Jane's Publishers, 1987), 62.

29. This is an average figure, based on NASA sources on the capabilities of its Haystack and Goldstone radars.

30. See Evan L. Schwartz, "The Looming Space Junk Crisis: It's Time to Take Out the Trash," *Wired*, June 2010.

31. Siddiqi, *Sputnik and the Soviet Space Challenge*, 174.

32. Pete Trabucco, "Shuttle Retirement: The End of an Era?" *Ad Astra* 23 (Winter 2011): 16.

33. On the *Apollo 1* accident, see Burrows, *This New Ocean*, 406–13.

34. In the *Challenger* incident, the rubber seals on one of the solid rocket boosters failed during a cold-weather launch, causing a fire and explosion that cost the lives of seven astronauts. In the later *Columbia* disaster, hard foam insulation falling during launch from the supercooled, liquid-fuel booster (not a problem with older Apollo capsules that had sat *atop* their Saturn V rockets) damaged protective tiles on the side-mounted shuttle *Columbia*, causing it to suffer a catastrophic structural failure upon reentry and killing all seven astronauts aboard.

35. On the *Salyut* and *Mir* programs, see Brian Harvey, *Russia in Space: The Failed Frontier?* (Chichester, UK: Springer, 2001), 14–16, and 22–28, respectively.

36. Trabucco, "Shuttle Retirement."

37. NASA, "International Space Station: Facts and Figures," http://www.nasa.gov/mission_pages/station/main/onthestation/facts_and_figures.html.

38. On the origins of this program, see Susan Eisenhower, ed., *Partners in Space: US-Russian Cooperation After the Cold War* (Washington, DC: Eisenhower Institute, 2004), 92.

39. Neufeld, *The Rocket and the Reich*, 61–63.

40. On the effects of these tests, see James Clay Moltz, *The Politics of Space Security: Strategic Restraint and the Pursuit of National Interests*, 2nd ed. (Stanford, CA: Stanford University Press, 2011), 99–100, 119–21.

41. Lt. Col. Clayton K. S. Chun, *Shooting Down a "Star": Program 437, the US Nuclear ASAT System, and Present-Day Copycat Killers* (Cadre Paper No. 6, Maxwell Air Force Base, AL, April 2006).

42. On these tests, see Johnson, *Soviet Military Strategy in Space*, table 6, 154.

43. For more on this test, see Moltz, *The Politics of Space Security*, 202.

44. On the test, see J.-C. Liou and N. L. Johnson, "Physical Properties of the Large Fengyun-1 Breakup Fragments," *Orbital Debris Quarterly* 12 (April 2008): 4.

45. See Joan Johnson-Freese, *Heavenly Ambitions: America's Quest to Dominate Space* (Philadelphia: University of Pennsylvania Press, 2009), 80–89.

46. For more details on these systems, see Phillip J. Baines, "Prospects for 'Non-Offensive' Defenses in Space," in James Clay Moltz, ed., *New Challenges in Missile Proliferation, Missile Defense, and Space Security* (Occasional Paper No. 12, Monterey, Calif.: Monterey Institute of International Studies, Center for Nonproliferation Studies, July 2003).

47. For an extended discussion of this comparison, see James A. Vedda, "Planes, Trains, Automobiles, and Spaceships," chapter 7 in *Becoming Spacefarers: Rescuing America's Space Program* (Online: Xlibris, 2012).

2. THE POLITICS OF THE SPACE AGE

1. William E. Burrows, *This New Ocean: The Story of the First Space Age* (New York: Random House, 1998), xi.
2. William H. Schauer, *The Politics of Space: A Comparison of the Soviet and American Programs* (New York: Holmes and Meier, 1976), 71.
3. For more on these negotiations, see Christopher C. Joyner, *Governing the Frozen Commons: The Antarctic Regime and Environmental Protection* (Columbia: University of South Carolina Press, 1998).
4. Quoted in Thanos P. Dokos, *Negotiations for a CTBT 1958–1994: Analysis and Evaluation of American Policy* (Lanham, MD: University Press of America, 1995), 3.
5. "Science: A Shot at the Moon," *Time*, March 10, 1958.
6. Lee A. DuBridge, "Plain Talk About Space Flight" (1958), in Lester M. Hirsch, ed., *Man and Space: A Controlled Research Reader* (New York: Pitman Publishing, 1966), 101.
7. John M. Logsdon, *John F. Kennedy and the Race to the Moon* (New York: Palgrave Macmillan, 2010), 1.
8. For more on early U.S.-Soviet space conflicts, see Walter A. McDougall, *The Heavens and the Earth: A Political History of the Space Age* (New York: Basic Books, 1985), esp. 263–360.
9. On this point, see Peter L. Hays, *United States Military Space Into the Twenty-First Century* (Colorado Springs: Institute for National Security Studies, September 2002), 79.
10. Bruce Berkowitz, "The National Reconnaissance Office at 50: A Brief History" (Chantilly, VA: Center for the Study of National Reconnaissance, September 2011).
11. Secretary of State Dean Rusk, "Address, June 16, 1962," *U.S. State Department Bulletin* (July 2, 1962), 5.
12. L. C. McHugh, "The Why of Space Programs" (1962), in Lester M. Hirsch, ed., *Man and Space: A Controlled Research Reader* (New York: Pitman Publishing, 1966), 152.
13. Ibid., 153.
14. On NASA's budget in these years, see Logsdon, *John F. Kennedy and the Race to the Moon*, 134–36.
15. Eric Sterner, "Five Myths About NASA," *Washington Post*, July 1, 2011.
16. On the N-1's problems, see Asif A. Siddiqi, *The Soviet Space Race with Apollo* (Gainesville: University Press of Florida, 2003), 546–64.
17. See the text of the Outer Space Treaty on the U.S. State Department website, http://www.state.gov/t/isn/5181.htm.
18. On this proposal and an earlier variant supported by the non-aligned movement, see Detlev Wolter, *Common Security in Outer Space and International Law* (Geneva: United Nations Institute for Disarmament Research, 2006), 175.

19. On the Apollo fire, see Burrows, *This New Ocean*, 406–413.

20. Siddiqi, *The Soviet Space Race with Apollo*, 576–90.

21. Ibid., 590.

22. Burrows, *This New Ocean*, 416–17.

23. See Siddiqi, *The Soviet Space Race with Apollo*, chapters 17 and 18.

24. On the early history of the French rocket program, see Gunther Seibert, "The History of Sounding Rockets and Their Contribution to European Space Research" (Noordwijk, Netherlands: ESA Publications Division, HSR-38, November 2006).

25. For more on Japan's early space program, see Saadia M. Pekkanen and Paul Kallender-Umezu, *In Defense of Japan: From the Market to the Military in Space Policy* (Stanford, CA: Stanford University Press, 2010).

26. The "5" denoted the first success after four failed attempts with the Lambda rocket.

27. Brian Harvey, *China's Space Program: From Conception to Manned Spaceflight* (New York: Springer-Praxis, 2004), 24–26.

28. On this period, see Roger Handberg and Zhen Li, *Chinese Space Policy: A Study in Domestic and International Politics* (New York: Routledge, 2007), 72–77.

29. See the text of the convention on the website of the UN Office for Outer Space Affairs, http://www.oosa.unvienna.org/oosa/SpaceLaw/liability.html.

30. Nicholas Johnson, *Soviet Military Strategy in Space* (London: Jane's Publishers, 1987), 62.

31. See the text of the convention on the website of the UN Office for Outer Space Affairs, http://www.oosa.unvienna.org/oosa/SORegister/regist.html.

32. On this process, see Kazuto Suzuki, *Policy Logics and Institutions of European Space Collaboration* (Aldershoot, UK: Ashgate, 2003).

33. Ibid.

34. Johnson, *Soviet Military Strategy in Space*, 146–47.

35. On the negotiation's history, see Wolter, *Common Security in Outer Space*, 89.

36. See the text of the treaty on the website of the UN Office for Outer Space Affairs, http://www.oosa.unvienna.org/oosa/SpaceLaw/moon.html.

37. Sundara Vadlamudi, "Indo-U.S. Space Cooperation: Poised for Take-Off?" *Nonproliferation Review* 12 (March 2005), 203.

38. On these talks, see Johnson, *Soviet Military Strategy in Space*, 158–61.

39. Ibid., 160.

40. For more on SDI, see Burrows, *This New Ocean*, 532–43.

41. Susan Eisenhower, ed., *Partners in Space: US-Russian Cooperation After the Cold War* (Washington, DC: Eisenhower Institute, 2004), 19.

42. Pat Norris, *Spies in the Sky: Surveillance Satellites in War and Peace* (New York: Springer-Praxis, 2008), 78.

43. Eisenhower, *Partners in Space*, 33–36.

44. Brian Harvey, *Russia in Space: The Failed Frontier?* (New York: Springer-Praxis, 2001), 281–88.

45. Nancy Gallagher, "International Cooperation and Space Governance Strategy," in Eligar Sadeh, ed., *Space Strategy in the 21st Century: Theory and Policy* (New York: Routledge, 2013), 61.

46. For more on these events, see Joan Johnson-Freese, *Space as a Strategic Asset* (New York: Columbia University Press, 2007), 143–44.

47. On the planning for *Shenzhou V*, see Gregory Kulacki and Jeffrey G. Lewis, *A Place for One's Mat: China's Space Program, 1956–2003* (Cambridge, MA: American Academy of Arts and Sciences, 2009), 26–29.

48. Moltz, *Asia's Space Race*.

49. On India's shift, see G. S. Sachdeva, "Space Policy and Strategy of India," in Eligar Sadeh, ed., *Space Strategy in the 21st Century: Theory and Policy* (New York: Routledge, 2013).

3. CIVIL SPACE: SCIENCE AND EXPLORATION

1. Arthur C. Clarke, *The Exploration of Space*, rev. ed. (New York: Harper and Brothers, 1959), 188.

2. Ibid., 188–89.

3. Walter A. McDougall, *The Heavens and the Earth: A Political History of the Space Age* (New York: Basic Books, 1985), 13.

4. Ibid., 380–86.

5. Jim Green, quoted in Marianne Dyson, "Visions and Voyages: Planetary Science Priorities for 2013–2022," *Ad Astra* 23 (Fall 2011): 31.

6. Roger Handberg and Zhen Li, *Chinese Space Policy: A Study in Domestic and International Politics* (New York: Routledge, 2007), 169.

7. Donald K. Yeomans, "Beware of Errant Asteroids," *New York Times*, February 10, 2013, 12.

8. Andrew E. Kramer, "After Assault from the Heavens, Russians Search for Clues and Count Blessings," *New York Times*, February 17, 2013, 6.

9. Pascale Ehrenfreund, Chris McKay, John D. Rummel, Bernard H. Foing, Clive R. Neal, Tanja Masson-Zwaan, Megan Ansdell et al., "Toward a Global Space Exploration Program: A Stepping Stone Approach," *Advances in Space Research* 49 (2012): 5.

10. Asif A. Siddiqi, *The Soviet Space Race with Apollo* (Gainesville: University Press of Florida, 2003), 793–94.

11. Text of the National Aeronautics and Space Act of 1958, http://history.nasa.gov/spaceact.html.

12. On NASA's early cooperative programs, see John M. Logsdon, ed. (with Dwayne A. Day and Roger D. Launius), *Exploring the Unknown: Selected Documents in the History of the U.S. Civilian Space Program*, vol. 2, *External Relationships* (Washington, DC: NASA, 1996).

13. Ibid., 9.

14. See Susan Eisenhower, ed., *Partners in Space: US-Russian Cooperation After the Cold War* (Washington, DC: Eisenhower Institute, 2005), 42.

15. Giles Alston, "Diplomacy in Orbit," *The World Today*, May 1997, 117.

16. Text of President Bush's speech at NASA, January 14, 2004, http://www.nasa .gov/pdf/54868main_bush_trans.pdf.

17. Matt Greenhouse, "The James Webb Telescope Mission" (lecture at the Naval Postgraduate School, Monterey, California, November 3, 2011).

18. Text of the National Space Policy of the United States of America, June 28, 2010, http://www.whitehouse.gov/sites/default/files/national_space_policy _6–28–10.pdf, 7.

19. Joan Johnson-Freese, *Space as a Strategic Asset* (New York: Columbia University Press, 2007), 248–51.

20. Ibid., 252.

21. On ESA's formation, see Kazuto Suzuki, *Policy Logics and Institutions of European Space Collaboration* (Aldershot, UK: Ashgate, 2003).

22. On this issue, see Johnson-Freese, *Space as a Strategic Asset*, 172.

23. Ibid., 173.

24. Ibid., 176.

25. On these trends, see Bertrand de Montluc, "Russia's Resurgence: Prospects for Space Policy and International Cooperation," *Space Policy* 26, no. 1 (February 2010).

26. Stephen Clark, "Russia: Computer Crash Doomed Russian Mars Probe," Space.com, February 10, 2012.

27. Peter B. de Selding, "ESA Ruling Council OKs Funding for Mars Mission with Russia," *Space News*, March 19, 2012, 1.

28. Ivan Safronov, "Plany Roskosmosa uspeshno zapyshcheny na orbitu" (Roscosmos Plans Successfully Launched Into Orbit), *Kommersant*, March 13, 2012.

29. On JAXA's formation, see Saadia M. Pekkanen and Paul Kallender-Umezu, *In Defense of Japan: From the Market to the Military in Space Policy* (Stanford, CA: Stanford University Press, 2010), 62–65.

30. On these missions, see James Clay Moltz, *Asia's Space Race: National Motivations, Regional Rivalries, and International Risks* (New York: Columbia University Press, 2012), 57–59.

31. On Japan's international aims, see Hirotaka Watanabe, "Japan's Space Strategy: Diplomatic and Security Challenges," in Eligar Sadeh, ed., *Space Strategy in the 21st Century: Theory and Policy* (New York: Routledge, 2013).

32. On China's space organizations, see Eric Hagt, "Emerging Grand Strategy for China's Defense Industry Reform," in Roy Kamphausen, David Lai, and Andrew Scobell, eds., *The PLA at Home and Abroad: Assessing the Operational Capabilities of the Chinese Military* (Carlisle, PA: Army War College, 2010).

33. On the history of China's human spaceflight program, see Gregory Kulacki and Jeffrey G. Lewis, *A Place for One's Mat: China's Space Program, 1956–2003* (Cambridge, MA: American Academy of Arts and Sciences, 2009), 19–29.

34. Handberg and Li, *Chinese Space Policy*, 150.

35. On the *Chang'e* program, see Patrick Besha, "Policy Making in China's Space Program: A History and Analysis of the *Chang'e* Lunar Orbiter Project," *Space Policy* 26, no. 4 (November 2010).

36. Brian Harvey, *China's Space Program: From Conception to Manned Spaceflight* (New York: Springer-Praxis, 2004), 160–64.

37. On these activities, see Moltz, *Asia's Space Race*, 91, 94–95.

38. On these points, see Handberg and Li, *Chinese Space Policy*, 170–73.

39. On this legacy, see Moltz, *Asia's Space Race*, 113.

40. Ajey Lele, "An Asian Moon Race?" *Space Policy* 26, no. 4 (November 2010): 223–24.

41. For more on India's early cooperative programs, see Sundara Vadlamudi, "Indo-U.S. Space Cooperation: Poised for Take-off?" *Nonproliferation Review* 12 (March 2005): 199–223.

42. Eisenhower, *Partners in Space*, 44.

43. Moltz, *Asia's Space Race*, 123.

44. Personal discussion with NASA official, Washington, DC, June 2010.

45. Reuters, "India to Launch Mission to Mars This Year, Says President," February 21, 2013.

46. Claude Lafleur, "Costs of US Piloted Programs," *Space Review*, March 8, 2010, http://www.thespacereview.com/article/1579/1.

47. "Europe's ISS Chief Endorses China Invite," *Space News*, December 19, 2011, 3.

48. John Logsdon, "Opening the Door for International Cooperation," *Space News*, September 27, 2010, 19.

49. Ibid.

50. Bret G. Drake, "Strategic Considerations of Human Exploration of Near-Earth Asteroids" (presentation at the 2012 IEEE Aerospace Conference, Big Sky, Montana, March 5, 2012).

51. See Clara Moskowitz, "Orbital Debris Problem Has Reached Tipping Point, Report Warns," *Space News*, September 5, 2011, 13.

52. Frank A. Rose (remarks at symposium "Sustainable Space Development and Utilization for Humankind," Tokyo, Japan, March 1, 2012, available at http://www.state.gov/t/avc/rls/184897).

53. See NASA, *Orbital Debris Quarterly News* 17, no. 1 (January 2013).

54. For a detailed discussion of these issues, see Michael Listner, "Revisiting the Liability Convention: Reflections on ROSAT, Orbital Space Debris, and the Future of Space Law," *Space Review*, October 17, 2011, http://www.thespace review.com/article/1948/1.

55. On these issues, see the National Center for Remote Sensing, Air, and Space Law, University of Mississippi School of Law, *Res Communis*, http://rescommunis.olemiss.edu/?s=liability.

56. Tom Jones, "Steps for Planetary Defense," *Ad Astra* 23 (Spring 2011): 16–19.

57. Pete Worden (remarks at the International Space Development Conference, Huntsville, Alabama, May 20, 2011).

58. Teller, quoted in Tad Friend, "Vermin of the Sky: Who Will Keep the Planet Safe from Asteroids?" *The New Yorker*, February 18, 2011, 27.

59. On recent protection work, see William J. Broad, "Vindication for Entrepreneurs Watching the Sky: Yes, It Can Fall," *New York Times*, February 17, 2013, 1.

60. On the original decision, see Eligar Sadeh and James P. Lester, "Space and the Environment," in Eligar Sadeh, ed., *Space Politics and Policy: An Evolutionary Perspective* (Boston: Kluwer Academic Publishers, 2002), 158.

61. Molly K. Macauley, "Environmentally Sustainable Human Space Activities: Can Challenges of Planetary Protection Be Reconciled?" *Astropolitics* 5 (September–December 2007): 216–17.

62. Saara Reiman, "Is Space an Environment?" *Space Policy* 25 (May 2009): 81–87.

63. See Neil deGrasse Tyson, "The Case for Space," *Foreign Affairs* 91 (March/April 2012): 22–33.

64. Andreas Diekmann (remarks presented at the International Space Development Conference, Huntsville, AL, May 19, 2011).

65. Michael Griffin (remarks presented at the National Space Symposium, Colorado Springs, April 13, 2011).

4. COMMERCIAL SPACE DEVELOPMENTS

1. Jesse McKinley, "Out of This World! Space, the Ultimate Getaway," *New York Times*, Travel, September 9, 2012, 1.

2. Space Foundation, *The Space Report: 2012* (Colorado Springs: Space Foundation, 2012), 32.

3. See, for example, the optimistic account in Neil McAleer, "Industry in Space," in *The Omni Space Almanac: A Complete Guide to the Space Age* (New York: World Almanac, 1987), esp. 272–76.

4. On these developments, see Stephen B. Johnson, "Space Business," in Eligar Sadeh, ed., *Space Politics and Policy: An Evolutionary Perspective*, 241–80 (Boston: Kluwer Academic Publishers, 2002), 258.

5. See Intelsat, "Our History," http://www.intelsat.com/about-us/history/.

6. Ibid.

7. Space Foundation, *The Space Report: 2012*, 32.

8. Ibid., 35.

9. NASA, "Landsat 1," http://landsat.gsfc.nasa.gov/about/landsat1.html.

10. Centre National d'Etudes Spatiales (CNES), "SPOT," http://www.cnes.fr/web/CNES-en/1415-spot.php.

11. Joan Johnson-Freese, *Space as a Strategic Asset* (New York: Columbia University Press, 2007), 38. The rest of this paragraph draws in part on Johnson-Freese's discussion of this issue.

12. Space Foundation, *The Space Report: 2011* (Colorado Springs: Space Foundation, 2011), 87.

13. The author thanks Karen Andersen for clarifying this point.

14. Space Foundation, *The Space Report: 2013*, 55.

15. "SpaceX's Falcon 1e Rocket Replaces Cheaper Falcon 1," *Space News*, August 10, 2009.

16. Space Foundation, *The Space Report: 2012*, 39.

17. "Looking Beyond WRC-12," editorial, *Space News*, March 12, 2012, 26.

18. For more on these requirements, see Nathan C. Goldman, "Space Law," in Eligar Sadeh, ed., *Space Politics and Policy: An Evolutionary Perspective*, 163–80 (Boston: Kluwer Academic Publishers, 2002), 170.

19. See U.S. House of Representatives, *Final Report of the Select Committee on U.S. National Security and Military/Commercial Concerns with the People's Republic of China* (unclassified version), Report 105–851 (May 1999).

20. See, for example, Alastair Iain Johnston, W. K. H. Panofsky, Marco Di Capua, and Lewis R. Franklin, *The Cox Committee Report: An Assessment*, ed. M. M. May (Stanford, CA: Center for International Security and Cooperation, Stanford University, December 1999).

21. Mike Gold, corporate counsel, Bigelow Aerospace (remarks at the National Space Forum 2008, Washington, DC, February 7, 2008).

22. See U.S. House of Representatives, *Export Controls, Arms Sales, and Reform: Balancing U.S. Interests, Part II*, hearing before the Foreign Affairs Committee, serial no. 112–127 (February 7, 2012).

23. Sara Sorcher, "U.S. Recommends Relaxing Satellite Export Controls," *National Journal*, April 19, 2012.

24. SES chief executive Romain Bausch, quoted in Peter B. de Selding, "Chinese Hardware, Financing Changing Satcom Landscape," *Space News*, January 21, 2013, 1.

25. Stephen Clark, "Obama Signs Law Easing Satellite Export Controls," Spaceflightnow.com, January 3, 2013.

26. See the website of Surrey Satellite Technology, Ltd., http://www.sstl.co.uk/.

27. Space Foundation, *The Space Report: 2011*, 135.

28. Ibid.

29. Space Foundation, *The Space Report: 2012*, 33.

30. Peter B. de Selding, "ESA Subsidy Boosts Arianespace Into Black," *Space News*, April 16, 2012, 28.

31. See http://www.spacex.com/.

32. Garrett Reisman (SpaceX's project manager for human spaceflight), "SpaceX Commercial Spaceflight" (lecture, Naval Postgraduate School, Monterey, Calif., March 15, 2013).

33. SpaceX, "Launch Manifest," http://www.spacex.com/launch_manifest.php.

34. Reisman, "SpaceX Commercial Spaceflight."

35. SpaceX, "Falcon Heavy," http://www.spacex.com/falcon_heavy.php.

36. Space Foundation, *The Space Report: 2012*, 77–79.

37. Clara Moskowitz, "Private Spaceflight Industry at Big Turning Point, Advocates Say," *Space News*, October 29, 2012, 14.

38. Dan Leone, "Virgin Galactic Granted License Exemption for Spaceflight Experience," *Space News*, April 16, 2012, 22.

39. Andrew Brearley, "Mining the Moon: Owning the Night Sky?" *Astropolitics* 4 (Spring 2006): 46.

40. Margaret Morris, *Moon Base and Beyond* (Detroit: Scribal Arts, 2013), 79.

41. On lunar settlement and resources, see Paul Spudis, "Lunar Resources: Unlocking the Space Frontier," *Ad Astra* 23 (Summer 2011): 12.

42. Ibid.

43. Greg Baiden, "Living Off the Land: Lunar Mining for Human Space Settlement," *Ad Astra* 23 (Summer 2011): 37.

44. Dan Leone, "Asteroid Mining Venture to Start with Small, Cheap Space Telescopes," *Space News*, April 30, 2012, 7.

45. Dan Leone, "Space-Based Resource Exploitation Still Decades Off, Experts Say," *Space News*, October 22, 2012, 13.

46. National Space Society (NSS), "Space Solar Power: Limitless Clean Energy from Space," summary report, available at http://www.nss.org/settlement/ssp/.

47. "Looking Beyond WRC-12."

48. K. K. Nair, *Space: The Frontiers of Modern Defence* (New Delhi, India: Knowledge World Press, 2006), 141.

49. Safa Haeri, "Cuba Blows the Whistle on Iranian Jamming," *Asia Times* online, August 22, 2003.

50. On these developments, see Space Foundation, *The Space Report: 2012*, 86–87.

51. Peter B. de Selding, "NASA Picks Loral for $230 Million Laser Com Demonstration," *Space News*, April 12, 2012, 12.

52. For more on these questions, see the website of the Secure World Foundation and the many resources listed on the "Space Situational Awareness" page, http://swfound.org/resource-library/space-situational-awareness/.

53. Peter B. de Selding, "Satellite Operators Solicit Bids to Create Orbital Database," *Space News*, November 23, 2009, 5.

54. Space Foundation, *The Space Report: 2012*, 105.

55. Bertrand de Montluc, "Russia's Resurgence: Prospects for Space Policy and International Cooperation," *Space Policy* 26, no. 1 (February 2010).

56. On this issue, see Michael Beckley, "China's Century? Why America's Edge Will Endure," *International Security* 36 (Winter 2011/12), 41–78.

57. This paragraph draws on Craig Covault, "Surging Chinese Espionage Targets U.S. Space Companies," *AmericaSpace*, Americaspace.org, February 27, 2012; and also John Walcott, "Chinese Espionage Campaign Targets U.S. Space Technology," *Bloomberg News*, April 18, 2012.

58. Space Foundation, *The Space Report: 2011*, 139.

59. "Looking Beyond WRC-12."

5. MILITARY SPACE: EXPANDED USES AND NEW RISKS

1. See, for example, Ajey Lele, "Should India Conduct an ASAT Test Now?" in Ajey Lele, ed., *Decoding the International Code of Conduct for Space Activities*, 155–57 (New Delhi, India: Pentagon Security International, 2012).
2. The U.S. Department of Defense defines the term "space force application" as "combat operations in, through, and from space to influence the course and outcome of conflict by holding terrestrial targets at risk." See U.S. Joint Chiefs of Staff, "Space Operations" (Joint Publication 3–14, January 6, 2009), II-10.
3. For a detailed discussion of the physics behind space-based observation technologies, see "Remote-Sensing Payloads," in Jerry Jon Sellers et al., *Understanding Space: An Introduction to Astronautics*, 2nd ed., 382–93 (Boston: McGraw-Hill, 2004).
4. On these capabilities, see Jeffrey T. Richelson, "Signals Intelligence," chapter 8 in *The U.S. Intelligence Community* (Boulder, CO: Westview Press, 2012).
5. See, for example, Steven Lambakis, *On the Edge of Earth: The Future of American Space Power* (Lexington: University Press of Kentucky, 2001).
6. See, for example, Michael Krepon (with Christopher Clary), *Space Assurance or Space Dominance? The Case Against Weaponizing Space* (Washington, DC: Henry L. Stimson Center, 2003).
7. David Wright, Laura Grego, and Lisbeth Gronlund, *The Physics of Space Security: A Reference Manual* (Cambridge, MA: American Academy of Arts and Sciences, 2005). The rest of this section draws considerably on this study.
8. Ibid., 8–11.
9. See Jeffrey T. Richelson, *America's Space Sentinels: DSP Satellites and National Security* (Lawrence: University Press of Kansas, 1999), 125; also Phillip J. Baines, "Prospects for 'Non-Offensive' Defenses in Space," in James Clay Moltz, ed., *New Challenges in Missile Proliferation, Missile Defense, and Space Security* (CNS Occasional Paper No. 12, Monterey Institute of International Studies, Center for Nonproliferation Studies, Monterey, CA, July 2003).
10. The author thanks Charlie Racoosin for suggesting this point.
11. Baines, "Prospects for 'Non-Offensive' Defenses in Space."
12. Space Foundation, *The Space Report: 2011* (Colorado Springs: Space Foundation, 2011), 43.
13. Ibid., 88.
14. One such system is the MIRACL laser, which was reportedly used in a test against a U.S. Air Force satellite in 1997. On this incident, see Steven Lambakis, *On the Edge of Earth: The Future of American Space Power* (Lexington: University Press of Kentucky, 2001), 258.
15. Laura Grego, "A History of Anti-Satellite Programs" (report from the Union of Concerned Scientists, January 2012), 10.
16. Ibid., 9.

17. Sharon Weinberger, "3 Theories About the Air Force's Mystery Space Plane, X-37B," *Popular Science*, May 1, 2012.

18. Ibid., 12.

19. "Military Wipes Out Iraqi GPS Jammers," FoxNews.com, March 25, 2003; also Paul Richter and Kim Murphy, "Evidence Cited of Russian Arms in Iraq," *Boston Globe*, January 10, 2004.

20. Futron, "INTEL: Global Military Space," *Milsat Magazine* 3 (September 2009).

21. Dean Cheng, "China's Military Role in Space," *Strategic Studies Quarterly* 6 (Spring 2012): 67.

22. Office of the Secretary of Defense, *Annual Report to Congress: Military and Security Developments Involving the People's Republic of China* (Washington, DC: U.S. Department of Defense, 2011), 37.

23. Barry D. Watts, written testimony for the U.S.-China Economic and Security Review Commission on the Implications of China's Military and Civil Space Programs (Capitol Building, Washington, D.C., May 11, 2011), 3–4.

24. Office of the Secretary of Defense, *Annual Report to Congress: Military and Security Developments Involving the People's Republic of China,* 35.

25. Watts, written testimony for the U.S.-China Economic and Security Review Commission, 9.

26. Ibid., 10.

27. Gordon G. Chang, "The Space Arms Race Begins," *Forbes*, November 5, 2009.

28. "Eastdawn to Distribute Pleiades Data in China," *Space News*, October 10, 2011, 9.

29. "Europe: Multi-National MUSIS Space Surveillance?" *Defense Industry Daily*, July 19, 2011.

30. Gen. James Cartwright (remarks at the 2011 National Space Symposium, Broadmoor Hotel, Colorado Springs, April 14, 2011).

31. On India's military space efforts, see James Clay Moltz, *Asia's Space Race: National Motivations, Regional Rivalries, and International Risks* (New York: Columbia University Press, 2012), chapter 4.

32. K. S. Jayaraman, "India's Space Cell Leverages ISRO Technology for Armed Forces," *Space News*, February 16, 2009, A4.

33. Author's interviews in Tokyo, Japan, April 2009.

34. Ibid.; see also Saadia M. Pekkanen and Paul Kallender-Umezu, *In Defense of Japan: From the Market to the Military in Space Policy* (Stanford, CA: Stanford University Press, 2010).

35. U.S. Air Force, *Counterspace Operations* (Air Force Doctrine Document 2–2.1, August 2, 2004), viii.

36. Text of the National Space Policy of the United States of America, June 28, 2010, http://www.whitehouse.gov/sites/default/files/national_space_policy _6-28-10.pdf, 1.

37. Ibid., 2.
38. Ibid., 9.
39. Ibid., 7.
40. U.S. Department of Defense and the Office of the Director of National Intelligence, "National Security Space Strategy (Unclassified Summary)," January 2011, 4.
41. Ibid., 10.
42. Ibid., 14.
43. Everett Carl Dolman, "New Frontiers, Old Realities," *Strategic Studies Quarterly* 6 (Spring 2012): 39–86.
44. Amb. Gregory L. Schulte and Audrey M. Schaffer, "Enhancing Security by Promoting Responsible Behavior in Space," *Strategic Studies Quarterly* 6 (Spring 2012): 9.
45. Ibid.
46. Ibid.

6. SPACE DIPLOMACY

1. See Article IX of the Outer Space Treaty, http://www.state.gov/t/isn/5181.htm.
2. On this debate, see Joan Johnson-Freese, *Heavenly Ambitions: America's Quest to Dominate Space* (Philadelphia: University of Pennsylvania Press, 2009), 108–116.
3. See Russell Hardin, "The Tragedy of the Commons," *Science* 162 (1968): 1243–48.
4. See, for example, Daniel Deudney, "The Sky Is the Limit: Global Closure, Outer Space Geopolitics, and Planetary Protection" (paper presented at the annual convention of the American Political Science Association, Washington, DC, September 2–5, 2010).
5. Michelle Flournoy and Shawn Birmley, "The Contested Commons," *Proceedings* 135 (July 2009).
6. Detlev Wolter, *Common Security in Outer Space and International Law* (Geneva: United Nations Institute for Disarmament Research, 2006), 57.
7. On this process, see Ross Liemer and Christopher F. Chyba, "A Verifiable Limited Test Ban for Anti-Satellite Weapons," *Washington Quarterly* 33, no. 3 (July 2010): 151–52.
8. For more on this issue, see James A. Vedda, *Becoming Spacefarers: Rescuing America's Space Program* (Online: Xlibris, 2012), 164–65.
9. Here and subsequently in this paragraph, see text of the draft treaty at http://www.fmprc.gov.cn/eng/wjb/zzjg/jks/kjfywj/t408357.htm.
10. See letter from the Permanent Representative of China and the Permanent Representative of the Russian Federation addressed to the Secretary-General

of the Conference on Disarmament transmitting answers to the principal questions and comments on the draft "Treaty on the Prevention of the Placement of Weapons in Outer Space and the Threat or Use of Force Against Outer Space Objects (PPWT)" (Document CD-1872, August 18, 2009).

11. Council of the European Union, "Draft Code of Conduct for Outer Space Activities" (Document 16560/08, Annex II, December 3, 2008).

12. See Michael Krepon, "Complaints About the Code," Armscontrolwonk.com, February 5, 2012. Krepon provides the examples of the 1972 Incidents at Sea (Richard Nixon), the 1989 Dangerous Military Practices Agreement (George H. W. Bush), and the 2002 Hague Code of Conduct Against Ballistic Missile Proliferation (George W. Bush).

13. Jeff Kueter, "Do We Need a Code of Conduct for Space? Considering Recent Developments in the Effort to Change Behavior in Space" (report from the George C. Marshall Center, Arlington, VA, February 2012), 3.

14. Statement by Russian ambassador Alexey Borodavkin, Conference on Disarmament, Geneva, Switzerland, June 5, 2012.

15. Wolter, *Common Security in Outer Space and International Law*, 185–95.

16. The White House, "National Security Space Strategy (Unclassified Summary)," January 2011, 5.

17. Statement by Indian ambassador Sujata Mehta, Conference on Disarmament, Geneva, Switzerland, June 5, 2012.

18. Statement by U.S. ambassador Laura E. Kennedy, Conference on Disarmament, Geneva, Switzerland, June 5, 2012.

19. Ibid.

20. Ibid.

21. Statement by Mr. Wang Qun, at the thematic debate on outer space at the First Committee of the UN General Assembly, October 17, 2011.

22. Statement by Russian ambassador Alexey Borodavkin, Conference on Disarmament, Geneva, Switzerland, June 5, 2012.

23. Admiral Mike Mullen, "Commentary: We Can't Do It Alone," *Honolulu Advertiser*, October 6, 2006.

7. TRENDS AND FUTURE OPTIONS

1. Pascale Ehrenfreund, Chris McKay, John D. Rummel, Bernard H. Foing, Clive R. Neal, Tanja Masson-Zwaan, Megan Ansdell et al., "Toward a Global Space Exploration Program: A Stepping Stone Approach," *Advances in Space Research* 49 (2012): 4.

2. On the draft strategy, see Victor Zaborsky, "Russia: From Space Programs to Space Strategy?" *Space News*, April 30, 2012, 17.

3. Everett C. Dolman, *Astropolitik: Classical Geopolitics in the Space Age* (London: Frank Cass, 2002).

4. Everett C. Dolman and Henry F. Cooper, Jr., "Increasing the Military Uses of Space," in Charles D. Lutes and Peter L. Hays, eds., *Toward a Theory of Spacepower: Selected Essays* (Washington, DC: National Defense University Press, 2011), 115.

5. Everett Carl Dolman, "Space Power and US Hegemony: Maintaining a Liberal World Order in the 21st Century," in John M. Logsdon and Gordon Adams, eds., *Space Weapons: Are They Needed?* (Washington, DC: Space Policy Institute, George Washington University, October 2003), 95.

6. John J. Klein, *Space Warfare: Strategy, Principles, and Policy* (New York: Routledge, 2006), 159.

7. Nancy Gallagher, "International Cooperation and Space Governance Strategy," in Eligar Sadeh, ed., *Space Strategy in the 21st Century: Theory and Policy* (New York: Routledge, 2013), 71.

8. On this concept, see Daniel Vasella, Jakob Kellenberger, Luzius Wasescha, Roger de Weck, and Martine Brunschwig Graf, *"Soft" Gouvernance* ("Soft" Governance) (Geneva, Switzerland: Geneva Foundation for Governance and Public Policy, April 2007).

9. Space Foundation, *The Space Report: 2012* (Colorado Springs: Space Foundation, 2012), 136.

10. P. Ehrenfreund and N. Peter, "Toward a Paradigm Shift in Managing Future Global Space Exploration Endeavors," *Space Policy* 25 (November 2009): 245.

11. Nancy Gallagher, "Towards a Reconsideration of the Rules for Space Security," in John M. Logsdon and Audrey M. Schaffer, eds., *Perspectives on Space Security* (Washington, DC: Space Policy Institute, George Washington University, December 2005), 35.

12. Detlev Wolter, *Common Security in Outer Space and International Law* (Geneva, Switzerland: UN Institute for Disarmament Research, 2006), 204.

13. Col. (Ret.) Pat Frakes, "A Commons View: Global Commons or Common Resource and Heritage to Preserve for Future Generations," *Army Space Journal* 9 (Fall 2010/Winter 2011): 39.

14. On similar proposals, see Ross Liemer and Christopher F. Chyba, "A Verifiable Limited Test Ban for Anti-Satellite Weapons," *Washington Quarterly* 33, no. 3 (July 2010); James Clay Moltz, "Breaking the Deadlock on Space Arms Control," *Arms Control Today* 32, no. 3 (April 2002).

15. Arthur C. Clarke, *The Exploration of Space* (New York: Harper, 1959), 181.

INDEX

A-4 rocket, 15–16
ABM Treaty. *See* Anti-Ballistic Missile (ABM) Treaty
accidents: collision of U.S. commercial satellite and dead Russian spacecraft (2009), 25, 84, 142; crash of nuclear-powered Soviet satellite in Canada (1978), 47; and fuel systems, 20, 26; and human spaceflight, 26, 44; liability issues, 47, 84, 99; and lunar missions, 26, 44, 45, 80; and planetary exploration, 72; space shuttle accidents, 26, 73, 197n34. *See also specific countries*
Africa, 32, 99, 175
Agreement on the Rescue of Astronauts, the Return of Astronauts, and the Return of Objects Launched into Outer Space (1968), 44–45, 99
Akiyama, Toyohiro, 98
Aldrin, Buzz, 43(table)
Altair lunar module, 67
Antarctic Treaty, 40
Anti-Ballistic Missile (ABM) Treaty (1972), 47, 153, 179; U.S. withdrawal from, 51, 53
Antrix Corporation, 106
AP-MCSTA. *See* Asia-Pacific Multilateral Cooperation in Space Technology and Applications
Apollo-Soyuz spacecraft docking (1975), 48, 65
Apollo spacecraft, 17, 26, 43(table), 44, 45, 197n34
APRSAF. *See* Asia-Pacific Regional Space Agency Forum

APSCO. *See* Asia-Pacific Space Cooperation Organization
Ares I, Ares V rockets, 67
Ariane launcher, 48, 70, 96, 105, 107 (table)
Arianspace, 105
Ariel I, 65
Aristarchus of Samos, 14
arms control: and ASAT (anti-satellite systems), 49–50, 123; and Cold War politics, 132; Defense and Space Talks, 50; deterrence, 47, 131, 143, 152, 189; lack of post–Cold War progress, 8, 52–53, 150–53, 158, 188; Missile Technology Control Regime (MTCR), 18–19; Prevention of an Arms Race in Outer Space (PAROS) resolution (1981), 151, 152, 157; Russo-Chinese treaty proposal for banning space-based weapons, 157–59; UN Conference on Disarmament (CD), 8, 53, 150, 152–53, 157–58, 162–63; U.S. intransigence/diffidence regarding treaty proposals, 8, 53, 152–54, 157–59, 161, 163–65. *See also* space security; treaties, conventions, and international agreements
Armstrong, Neil, 43(table)
ASAT (anti-satellite) systems: absence of clear rules on testing and deployment, 8, 147–48, 184; and anti-ballistic missile defenses, 129; ASAT test of 2007 (China), 8, 29–30, 54, 122, 138, 139, 147–48, 184; Chinese capabilities, 135; and creation of orbital debris, 8; draft treaties on, 50, 157–59; Japanese capabilities, 139; need for ban on testing, 30,